Political Theory and Global Climate Change

Political Theory and Global Climate Change

edited by Steve Vanderheiden

The MIT Press
Cambridge, Massachusetts
London, England

For information about special quantity discounts, please e-mail special_sales@mitpress.mit.edu

This book was set in Sabon by SNP Best-set Typesetter Ltd., Hong Kong.

Printed on recycled paper and bound in the United States of America.

Library of Congress Cataloging-in-Publication Data

Political theory and global climate change / edited by Steve Vanderheiden.
    p.   cm.
    Includes bibliographical references and index.
    ISBN 978-0-262-22084-2 (hardcover : alk. paper)—ISBN 978-0-262-72052-6 (pbk. : alk. paper)
    1. Climatic changes—Political aspects.   2. Climatic changes—Social aspects.
3. Environmental policy.   4. Environmental justices.   5. Environmental ethics.
I. Vanderheiden, Steve.
    QC981.8.C5P657 2008
    363.738′74—dc22
                                                                          2008017047

10 9 8 7 6 5 4 3 2 1

Contents

Foreword   vii
John Barry

Introduction   xi
Steve Vanderheiden

Contributors   xxv

I Justice, Ethics, and Global Climate Change   1

1 Allocating the Global Commons: Theory and Practice   3
Leigh Raymond

2 A Perfect Moral Storm: Climate Change, Intergenerational Ethics, and the Problem of Corruption   25
Stephen Gardiner

3 Climate Change, Environmental Rights, and Emission Shares   43
Steve Vanderheiden

4 Environmental (In)justice in Climate Change   67
Martin J. Adamian

II Climate Change, Nature, and Society   89

5 Climate Change and Arctic Cases: A Normative Exploration of Social-Ecological System Analysis   91
Amy Lauren Lovecraft

6 Climatologies as Social Critique: The Social Construction/Creation of Global Warming, Global Dimming, and Global Cooling   121
Timothy W. Luke

7 Urban Sprawl, Climate Change, Oil Depletion, and Eco-Marxism   153
George A. Gonzalez

8 In the Wake of Katrina: Climate Change and the Coming Crisis of Displacement   177
Peter F. Cannavò

References   201
Index   225

Foreword

John Barry

With documentaries about climate change gaining not one but two Oscars in 2007, we can say something is happening in the public mind and in our popular culture with respect to the biggest environmental threat we face as a species. It is fitting that a book such as this follows in the wake of former U.S. Vice President Al Gore's widely acclaimed *An Inconvenient Truth*, its receipt of an Oscar, and the awarding of the Nobel Prize jointly to Al Gore and the Intergovernmental Panel on Climate Change (IPCC). Equally significant was the publication in November 2007 in Paris of the *Fourth Assessment Report* of the Intergovernmental Panel on Climate Change (IPCC) which gained much media attention following the Stern Report published in the autumn of 2006 by the UK Treasury. While the ideological attacks from "climate skeptics" have not and will not simply go away, there is a discernible movement within climate change politics (among some governments, political parties, citizens' groups, and policymakers) toward formulating solutions and toward action—increasingly wrapped in the new "green nationalism" of "energy security" and heightened concerns about the dire effects of climate change, leading to more environmental refugees in the world than people displaced from international and civil wars. Thus environmental/green concerns have become mainstreamed within conventional political, economic, and military institutions and modes of thinking and acting. With climate change, green issues—once so marginal to political debate—have arrived at the center of geopolitics.

Climate change is shaping and will continue to shape the political, economic, and cultural landscape as much as the biophysical landscape throughout this century. Alongside scientific information and technological innovation (and not just Western forms of scientific knowledge), we also need normative analyses in relation to the political and ethical implications of climate change. As with other forms of change, we need to think about the distributive impacts of such change: Who are the winners and losers? What are the appropriate forms of justice (ecological, global, social, and intergenerational) we need to deploy to help us think through the ethical dimensions of climate change? What sort of societies and economies can "best fit" and adapt to climate change?

This timely volume features a representative sample of the innovative and integrative scholarship within the broad field of environmental political theory/green political theory. It is perhaps a mark of the increasingly abstract character of most (Western) contemporary political theory that it seems to continue to proceed blissfully ignorant of the looming, multifaceted, and potentially catastrophic threat of climate change. Pick up any of the leading journals in the field of political theory and one finds the comforting debates about and within post-Rawlsian liberalism, "governance" and the relationship between state, civil society, and market, Habermasian discourse ethics, rereadings and reinterpretations of the "dead white men" of the Western political theory canon, and so on. This is not to belittle these scholarly efforts. But there is something more than a little odd (if not to say seriously wrong) with the field of political theory the vast bulk of which seems unaware or uninterested in the massive, global, and unprecedented socioecological changes that are happening now and will continue to shape the world throughout the twenty-first century. In particular, this abstraction is especially marked in UK and some European forms of political theorizing, the intellectual contexts of which (and associated scholarly institutional infrastructures) are lacking the integrative, interdisciplinary features that mark North American, Australian, and non-Western forms of scholarship, teaching, and research on political theory. This volume is an excellent example of the type of innovative, interdisciplinary scholarship that is increasingly needed to forward our thinking on climate change.

While the field of environmental/green political theorizing has matured in the last decade, it is still remarkable how little the concerns raised in this field (ranging from issues of animal welfare, critiques of dominant economic theory, and intellectual property rights in genetic material, to anthropogenic climate change and concerns with the contours, principles, and strategies of creating sustainable societies) have initiated a dialogue within mainstream political theory. This exchange is of course happening to some degree; even the mainstream political theory journals now contain articles on some of these "green" issues. But it is evident that even with the overwhelming scientific evidence that the burning of fossil fuels is dramatically altering the earth's climate with potentially devastating effects (especially to the most vulnerable human and nonhuman communities), there is still a lot more work to be done in political theory to address the issue of climate change.

This book presents a challenge to mainstream political theorizing even as it uses some of the issues, concepts, and debates within the liberal/distributive justice framework that characterizes most contemporary political theorizing. Part I on "Justice, Ethics, and Global Climate Change" looks at the distributive justice concerns raised by climate change to help us think through issues of justice associated with climate change. If we accept the dominant view that each member of the human species has an equal share of the capacity of the earth to absorb carbon emissions, what ethical and policy proposals flow from the "inconvenient truth" that a minority of the world's population (mostly those living in wealthy, industrialized countries) are not only using up a disproportionate (and therefore unfair) amount of this resource but are also the major cause and beneficiary of that unfairness? What of the relationship between concerns of global distributive and environmental injustice and obligations of justice to those yet to be born?

Part II on "Climate Change, Nature, and Society" shifts the issue away from a focus on "justice" toward a collection of interrelated concerns that we need to attend to in order to ground our thinking in some salient interdisciplinary, conceptual, and practical political/policy concerns around climate change. What does political theory look like when it is explicitly grounded in the material "realities" of socioecological processes and consequences? What are the ideological and knowledge/power strategies used and abused by protagonists in the political battle and drama for the hearts and minds of citizens, scientists, policymakers, and others? Given the inevitability of not being able to "solve" or "prevent" climate change, what forms of political and economic adaptation are needed or legitimate or expedient to deploy as "coping mechanisms" to help us (if we're lucky) to "muddle through"?

Climate change will shape not only the physical but the intellectual landscape of the twenty-first century, and green/environmental theorists will point the way in sketching out the contours of this new terrain for others to follow. Hence I see the contributors to this book as pioneers.

I applaud their moral and intellectual courage in facing what is an unprecedented ethical and material crisis, and I wish them well on their journey. Albert Einstein is credited with noting that the thinking that causes a problem is inadequate to solve the problem. The innovative thinking contained in this volume is precisely the type of new thinking we need to cope with the political and ethical consequences of the climate change crisis.

Introduction

Steve Vanderheiden

On the November 2007 release of the fourth and final assessment report of the Intergovernmental Panel on Climate Change (IPCC), the scientist, economist, and panel head Rajendra Pachauri declared: "If there's no action before 2012, that's too late, there is not time. What we do in the next two, three years will determine our future. This is the defining moment."[1] Echoing earlier reports but now calling the science establishing the existence and anthropogenic causes of climate change "unequivocal," the report identifies various mitigation strategies for policymakers to consider, warning that "unmitigated climate change would, in the long term, be likely to exceed the capacity of natural, managed, and human systems to adapt" and that "delayed emissions reductions significantly constrain the opportunities to achieve lower stabilization levels and increase the risk of more severe climate change impacts."[2] United Nations Secretary General Ban Ki-Moon, presiding over the release of the report, called on the United States and China—the two largest greenhouse gas (GHG) polluters—to play "a more constructive role" in future global climate policy negotiations, implicitly rebuking the George W. Bush administration for its 2001 formal withdrawal from the unratified Kyoto Protocol and undermining of the formal UN climate policy process with its 2006 Asia-Pacific Partnership on Clean Development and Climate.[3] As the world looks forward to a post-Kyoto agreement, given that treaty's expiration in 2012, the obstructionism that has characterized the U.S. role within the global climate policy process since 1997 could give way to genuine leadership, but only if the United States takes seriously its occasionally announced but rarely followed commitment to developing a fair and effective global climate policy.[4]

Such a commitment to fairness can be found in the 1992 UN Framework Convention on Climate Change (UNFCCC), which declared anthropogenic climate change to be a "common concern of mankind" and pledged concerted international action to prevent "dangerous anthropogenic interference with the climate system." The 192 signatory nations to that treaty acknowledged that "the largest share of historical and current global emissions of greenhouse gases has originated in developed countries" and resolved to design and empower a global climate regime in order to "protect the climate system for the benefit of present and future generations of mankind, on the basis of equity and in accordance with their common but differentiated responsibilities and respective capacities."[5] With the 1997 Kyoto Protocol, the first incarnation of this climate regime attempted to instantiate these normative commitments in global climate policy, assigning to the world's industrialized nations an average reduction of 5 percent from the 1990 baseline levels at which they had pledged to freeze their GHG emissions under the UNFCCC, to be achieved by the compliance period of 2008–2012. Citing the declared commitments to "equity" and the "differentiated responsibilities and respective capacities" of nations, developing countries like China and India were exempted from the first round of mandatory emission caps, given their relative poverty and significantly lesser historical and current causal responsibility for the problem, measured in terms of per capita GHG emissions.

Although the treaty has been in force since February 2005, having been ratified by 176 nations representing 63.7 percent of global GHG emissions, the United States continues to hold out, citing concerns about developing-country exemptions in justifying its rejection of the protocol's modest preliminary effort at realizing the UNFCCC's mandate and undermining the consensus that had earlier developed around the imperative that industrialized nations "take the lead" in reducing GHG emissions by agreeing to phase in caps on developing-country emissions later. As the Kyoto Protocol's 2012 expiration approaches, two unfortunate and related conclusions have become clear: that no global climate regime can be effective without the cooperation of the world's largest greenhouse polluter, and that the need for effective climate policy is no less urgent in 2008 than in 1992. Insofar as concerns about the protocol's fairness have served as a pretext for U.S. defection from the global regime, the predicted consequences of ongoing inaction underscore the importance of inquiries into fairness as applied to global climate policy. Absent some new normative consensus about the fairest way to proceed— a value judgment based in but distinct from the scientific facts about climate change—free riding by the nation responsible for over 22 percent of global GHG emissions will be sufficient to undermine the cooperative scheme.

As atmospheric GHG concentrations continue their upward trajectory unabated by recent policy efforts and the declared goal of avoiding "dangerous anthropogenic interference" with the earth's climate system becomes increasingly difficult to realize, the international stalemate over how to allocate climate-related costs has claimed fairness itself as among its victims. Having failed to instantiate their declared ideals in policy, the world's nations have made some harmful level of climate change inevitable. Because rich countries like the United States have refused to commit to meaningful action unless poor countries are also required to accept mandatory emission caps in the initial phase, the world's rich have allowed the very activity of their affluence (manifest as GHG emissions) to harm the world's disadvantaged. As the IPCC predicts of the likely effects of the unconstrained fossil fuel combustion and deforestation that continue to cause the problem while some of those nations most responsible for this outcome stubbornly resist acting to mitigate it, impacts "will fall disproportionately upon developing countries and the poor persons within all countries, and thereby exacerbate inequities in health status and access to adequate food, clean water, and other resources."[6]

At the core of this international impasse over how to assign the necessary costs associated with climate change mitigation is a conflict over what is required of the normative commitments aptly declared in the UNFCCC. Debate within the United States over the Kyoto Protocol, when it is not calling into question the integrity of the science on which the phenomenon is premised, has centered on what its fair share of climate-related burdens might be. Opponents have alleged (in the 1997 Byrd-Hagel Resolution, passed 95–0, threatening Senate rejection of the treaty as adopted at Kyoto) unfair "disparity of treatment" in assigning binding caps to the United States but not China or India and implied an unfair assignment of burdens in noting the "serious harm to the United States economy" that might result from such caps. Defenders of the "common but differentiated responsibilities" framework point out in reply that per capita emissions in developing countries are only a small fraction of those in the United States and that similar caps placed on 1990 baseline national emissions would effectively prohibit the further economic development of nations like India and China. Issues concerning each nation's fair share of responsibilities for climate change mitigation now constitute the primary obstacle to an effective global climate policy, highlighting the critical importance of normative analysis of competing claims about the requirements of fairness.

Anthropogenic climate change involves highly complex causal chains and is expected to produce consequences that are extraordinarily difficult to forecast, yet perhaps the most confounding aspects of the problem are political rather than scientific. With four scrupulously researched, meticulously prepared, and widely disseminated IPCC assessment reports published since 1990, we now know what causes climate change and can make reasonable estimates regarding its effects, and we know in general terms what needs to be done in order to minimize its most harmful consequences (i.e., reduce GHG emissions and maintain carbon sinks). Humans have aptly recognized climate change to involve issues of global justice, since it essentially entails a massive negative environmental externality created by the world's affluent to be disproportionately borne by those least responsible for it among the poor and future generations, and the world's nations have rightly identified ideals of fairness to serve as normative guides to action. But the value-based and political obstacles to effective global climate policy now far surpass the scientific and technical ones in importance. Compared to the intellectual resources devoted to the scientific study of climate change over the past two decades, relatively little attention has been paid to the normative political issues surrounding this uniquely global and thus far intractable environmental problem. This book aims to provide a partial remedy for that deficit. Many other fundamental social and political questions cannot be addressed within its pages, but it is a start.

While climate scientists must continue to sort out the facts surrounding the causes and likely effects of climate change, and while opponents of meaningful efforts to reduce those causes and mitigate those effects will surely continue to contest those facts in an effort to delay effective action, those trained in scholarly disciplines specializing in the critical assessment of value claims must not neglect aspects of global solutions that are not reducible to facts alone. Problems of designing and implementing an effective global response to climate change that promotes equity and is based in the "common but differentiated responsibilities" of nations are inherently normative and political, so political theorists should be uniquely well equipped to address various dimensions of those problems.

Climate change challenges our existing political institutions, ethical theories, and ways of conceptualizing the human relationship with the environment. It defies current principles of distribution, transcends current discourses on rights, and disrupts the sense of place on which our connections to the world are based. We desperately need to think more clearly about how best to understand the social and political obstacles to fair and effective global climate policy—the obstacles that are at the root of the current international climate policy impasse—and the conceptual tools of political theory can assist in this regard. The analyses and arguments in the following chapters depend on facts and insights gleaned from other areas of knowledge, and political theory's contribution to the debate surrounding appropriate global responses to climate change is but one contribution of many. But it is an important contribution.

Global climate change offers a unique case study for observers from a variety of backgrounds, both scholarly and otherwise, since it presents what most can agree are a set of problems that cut across a wide range of disciplines. Atmospheric scientists see a host of technical challenges in developing models that accurately predict effects on weather patterns of increasing atmospheric concentrations of GHGs as well as a further challenge in disseminating their findings to policymakers and the lay public in a manner that is comprehensible and impervious to distortion or manipulation. Economists see a global negative externality, where the costs of industrialization and affluence in some countries are being displaced onto others rather than being captured within the transaction costs of those activities, and must also grapple with manifold uncertainties in building climate impact and mitigation costs into forecasting models. Empirical political scientists see shortcomings in global institutional capacity for regulating GHG emissions or establishing carbon markets and must overcome numerous obstacles in translating impact predictions from climate scientists and economists into meaningful assessments about the social and political effects of climatic instability. Some climate skeptics, including the novelist Michael Crichton, who was called as the lead witness in a 2005 Senate Environment and Public Works Committee hearing on climate change, see an elaborate hoax perpetrated by environmental groups and dogmatic scientists designed to make the United States submit to an insidious world government.[7] Political theorists, viewing not only global climate change itself, but also the way it is seen by these and other observers, notice a variety of problems that are related to but distinct from those noted above, reflecting the variety of methods and conceptual lenses employed within the subfield. The chapters in this book illustrate that diversity.

With access to a plethora of illuminating methods and concepts—the following chapters include applications of analytic philosophy, conceptual analysis, critical theory, constitutional and legal theory, place-based value theories, neo-Marxism, and Critical Legal Studies—the political theorists contributing to this volume do not merely use the climate case as a mirror, reflecting back through the issue's analysis the various norms, conventions, and assumptions brought to bear on it. Instead they use the case as a kind of medium in which critical insights might be cultivated, and through which standard normative and empirical premises may be tested. Indeed, as the book's eight chapters demonstrate, global climate change cannot merely be regarded as an environmental problem of unprecedented proportions, but it is also, and perhaps equally, a problem for politics and society, and therefore for the norms and concepts through which we interpret our world, and on which we construct our social and political institutions.

Anthropogenic climate change forces us to rethink the nature of sovereignty, cosmopolitan justice, and the value of place; it requires us to revisit dichotomies that have long been established between human settlements and nature and to reconsider the causal relationships between human and ecological systems; and it makes us reexamine the role of concepts, norms, and values in creating environmental conditions under which others are widely expected to suffer. We must do these things because global climatic instability poses ecological and environmental threats, but also because it challenges the ideas, ideals, and institutions on which our world depends, and that are essential if we are to minimize those threats and sustain a livable and morally decent world. Normative political theories must be able to adapt to a changing world—indeed, a world that they have some role in changing and whose prior changes have shaped their development—and the discipline of political theory can assist in that theoretical adaptation, which is essential to our adaptation to the world. To fully understand the multifarious problems associated with global climate change, it is imperative that we understand the complex relationship between the animating norms, concepts, and theories engaged here and the phenomenon of which they are both cause and consequence, and appreciate the potential contributions that political theory (and political theorists) can make to mounting an appropriate response to the sort of problems climate change entails.

Environmental political theory, as an emerging subfield within political science, speaks with a plurality of voices. Elsewhere, I have enlisted one approach to normative theorizing about the environment in aiming to comprehend the nature of the problem of global climate change and to recommend a response to it,[8] but here my aim is different. If humanity is to meet the challenge of responding to climate change without abandoning our noblest ideals and most esteemed and worthy capacities or regressing to our worst ones, we must first understand it, in its full causes and effects. These causes and effects include, but are not limited to, those described in analyses of climate change by atmospheric scientists, economists, and empirical political scientists (among others); they also include the concepts, ideals, and norms in which our political theories are grounded and through which political theory seeks knowledge and understanding. In enlisting environmental political theory to interrogate the nature of the problem of global climate change, depth and breadth yield different but equally valuable insights. Here, I hope not only to illustrate the relevance of my preferred theoretical methodology to climate change, but to show how the plurality of methods, concepts, and approaches that comprise environmental political theory can illuminate a more complete set of problems and point to needed solutions. Appreciating the problem of global climate change requires an appreciation of myriad related problems, and my aim here is to enlist diverse voices so as to provide that background.

Plan of the Book

This anthology includes eight chapters organized around two major themes ("justice and ethics" and "nature and society"). In chapter 1, Leigh Raymond examines five arguments for the allocation of the global atmospheric commons commonly made in normative debates about global policy responses to climate change, critically examining the case for each. Relying on past cases of appropriation or allocation of other unclaimed resources from the "global commons"—of Antarctica, the oceans, and the moon—Raymond finds little precedent for any of the five standard allocation arguments. Instead the recurring Humean claim to exclusive national property rights based in possession (like those implicit in GHG emission rights) is often opposed by "a more radical, egalitarian rejection of any exclusive control over the earth's common resources that does not benefit all citizens of the world." Such a view can be seen, he suggests, in the Common Heritage of Mankind (CHM) principle that has been proposed for the management of the high seas and that is reflected in the Moon Treaty. This principled resistance to what Raymond terms the "enclosure" of the global commons contrasts with schemes that assume private-property-right allocations to be a necessary mechanism for avoiding the "tragedy of the commons" of an overappropriated atmosphere. Despite its explicit rejection in principle of the private allocation of the atmosphere's absorptive capacity, Raymond identifies several conceptual links between the CHM idea and the "contraction and convergence" proposal for an equal per capita assignment of national emissions shares, and sees in this ideal the potential to overcome several prominent normative objections to the privatization of the atmosphere.

In chapter 2, Stephen Gardiner aptly observes that "we cannot get very far in discussing why climate change is a problem without invoking ethical considerations," going on to discuss the unique and theoretically challenging nature of several of those considerations. Drawing on the idea of a "perfect storm" as an "unusual convergence of independently harmful factors where this convergence is likely to result in substantial, and possibly catastrophic, negative outcomes," Gardiner identifies three distinct moral "storms" that converge in the phenomenon of global climate change. In what he terms the "global storm" of widely dispersed causes and effects, the spatial fragmentation of agency undermines global efforts to curb GHG emissions, while in the "intergenerational storm" of delayed effects, agency is temporally fragmented, creating an intergenerational collective-action problem that defies straightforward solutions. Finally, Gardiner argues, the "theoretical storm" of conceptual confusion arising from several intersecting problems that challenge conventional terms of ethical analysis leads (in a "perfect storm") to what he terms the "problem of corruption." Observers may be tempted to selectively focus on some but not all moral challenges posed by the complexity of climate ethics, he argues, favoring their own roles in the crisis but to the detriment of a proper understanding of its manifold issues. Gardiner worries that this complexity may "turn out to be *perfectly convenient* for us, the current generation," leading us toward theoretically indefensible inaction rather than meaningful policy action, as we unjustifiably exploit our generational advantage over those who come later, and who might suffer climate-related harm as the direct result of our current choices.

In chapter 3, Steve Vanderheiden considers the roles potentially played by three environmental rights in shaping the design of a global climate policy regime: the rights (1) to develop; (2) to sustain a minimum level of per capita GHG emissions (termed "survival emissions"); and (3) to achieve climatic stability. Although none of these rights has yet been fully realized within international law, they each represent common normative claims within climate policy debates, and each has some foundation in both recognized universal human rights and political philosophy. Vanderheiden argues that these three rights, taken in combination, form the basis for a fair and effective global climate regime, and effectively model the primary normative constraints surrounding the way global GHG abatement efforts are assigned among nations and peoples. Crucially, he suggests, this combination of rights-claims illustrates the complexity of the distributive justice problem that lies at the root of the global allocation of atmospheric space, and the various issues of justice that such an allocation must consider. Moreover, he argues that these three kinds of environmental rights can be rank-ordered in a hierarchy of decreasingly basic claims (drawing on Henry Shue's work on basic rights[9]), resolving conflicts between the various rights-claims based in a lexical priority system and creating a foundation for the application of principles of cosmopolitan and intergenerational justice to the design of a global response to climate change mitigation and adaptation.

In chapter 4, Martin J. Adamian applies several insights from Critical Legal Studies (CLS) to problems in climate change politics and policy, calling into question the very possibility of using international environmental law to advance the ideals of justice expressed in the Framework Convention on Climate Change. Adamian notes that the liberal conception of justice, which typically focuses on issues of distribution, may not effectively capture the various dimensions of injustice present in global environmental politics, especially within the context of climate change. Moreover, he suggests, the aims of justice, when construed more broadly, may be frustrated by the institutional norms and incentives of international politics, as CLS critics have elsewhere alleged. Insofar as law reflects existing hierarchies of power and structures of domination, as the CLS critique alleges, one might expect international environmental law not to work in practice toward guaranteeing justice to the world's disadvantaged—an aim that, as Adamian suggests, is essential to fully addressing the problems of global climate change—but rather toward reinforcing those hierarchies and further entrenching that domination. He examines the system of international law on which the developing climate regime is based, and notes several areas in which it falls short of the liberal model of a neutral and effective legal system based on the rule of law, arguing that these shortcomings help explain why the climate regime continues to be frustrated by U.S. nonparticipation and obstructionism as well as by a set of systematic biases against some of the

world's most vulnerable peoples. Ultimately endorsing "greater democratization of environmental governance" as a partial remedy for these
problems inherent in mainstream liberal approaches to climate policy
development, Adamian's application of several CLS insights sounds a
cautionary note regarding the liberal approaches endorsed by Gardiner
and Vanderheiden.

In chapter 5 (commencing the volume's second thematic section, on
issues in "nature and society"), Amy Lovecraft examines the effect of
climate change on Arctic ecosystems through the use of social-ecological
systems (or SES) analysis. Using the Habermasian concept of the *lifeworld*—which contains "shared common understandings that individuals and social groups develop over time in their cultures, societies, and
personalities"—Lovecraft considers how global climate change has
already altered the lifeworlds of Arctic peoples, and how it might continue to do so. Social and political systems, particularly those in the
high-latitude areas most vulnerable to climatic instability, are linked to
ecological systems in manifold ways, she argues, and SES analysis can
help project how rapid changes in the latter may affect the former.
Taking normative standards like robustness or resilience as aims in SES
management, one can theorize not only the ecological responses of such
communities to changes in climate, but also the social consequences
likely to follow from resulting declines in ecosystem services. Cases of
fire management and sea-ice coverage illustrate how these aims might
likewise be applied to climate change, which is a macroscale environmental problem on which these smaller-scale management issues depend,
both causally and conceptually. At issue in adaptation within these
vulnerable Arctic systems, Lovecraft suggests, is "the ability to form and
implement long-range planning capacity to research and make decisions
related to climate change" through the use of analytic techniques and
their associated norms. Her chapter illustrates both the potential value
and difficulties in wielding SES methodologies in both theory and practice related to climate changes.

In chapter 6, Timothy W. Luke critically examines the social construction of "global warming" (along with "global cooling" and "global
dimming"), which depend not only on the noted effects of increasing
atmospheric concentrations of greenhouse gases, but also on the relationship between climate change and the human societies that it alters and
the various forms of social critique that it entails. Global dimming refers
to measurable decreases in solar radiation reaching the earth's surface
due to increasing concentrations of chemicals and particulates in the

atmosphere, which reflect this radiation back into space rather than allowing it to penetrate the atmosphere. These various anthropogenic phenomena result from humans fundamentally reshaping their environment, with the outcome, Luke suggests, that "human and natural life forms begin to inhabit a nature, which as habitat is being recreated by the output of corporate labs, major industries, and big agribusiness," with human and natural environments fused together into a new hybrid "urbanatura" that we must first understand before we can fruitfully address. These changes require us to rethink the nature of our environment, which he suggests we must now see "as an essentially built environment, a human-machine hybrid, or a vast artifice ironically fabricated by wastes, by-products, or effluents." Deconstructing the claims made by climate scientists as well as skeptics, Luke argues that "climatology as social critique reveals how starkly material inequalities express themselves tangibly at every point of sale and site of production," and forces us to move beyond conventional spatial concerns to new temporal ones. Like Gardiner, Luke notes that the causal mechanism of climate change affects others through time as well as across spatial borders, challenging received norms and reconstructing nature in terms with which it was not previously associated, but with which—largely as the result of anthropogenic climate change—it is now indelibly connected.

In chapter 7, George Gonzalez examines the relationship between urban sprawl, fossil fuel combustion, and climate change, and does so through the lens of a neo-Marxist political economy. As Gonzalez notes, sprawl is less the by-product of unplanned urban and suburban expansion than the consequence of a deliberate economic policy designed to increase consumption, and has historically relied on abundant supplies of cheap oil along with the presumption that its combustion was benign. Increasing global demand for petroleum combined with flagging supplies have made sprawl more economically costly (as commuters face increasing prices at the pump), while global environmental problems like climate change betray the inaccuracy of the second assumption, because it has increased the reliance on the personal automobile for transportation and dramatically increased the distance annually driven (and pollution created) by suburban and exurban commuters. Approaching global climate change causation from the perspective of value, Gonzalez suggests that Marx's concept of exchange value is partly responsible for the undervaluation of natural resources like oil, and thus has contributed to both urban sprawl and climate change. Exploring the histories of U.S. oil policy and the Federal Housing Authority, he suggests that the two combined to promote suburbanization and sprawl in order to spur demand for consumer durables, illustrating "Marx's contention that within capitalism economic demand is shaped to maximize the realization of profit." Pro-sprawl policies have specifically intended to increase oil consumption, Gonzalez suggests, and consistently eschewed conservation efforts that might have reduced per capita emissions of greenhouse gases in the United States, making the conversion to more climate-friendly transit options more tenable. At the root of this policy is a mistake in value theory, he argues, that the neo-Marxist critique of capitalism has long maintained, and that analysts of climate politics would do well to acknowledge.

In chapter 8, Peter F. Cannavò takes on the issue of adaptation to climatic changes, which is often understood to comprise part of the charge of a global climate regime that focuses primarily on mitigation. In response to those who, like Wilfred Beckerman and Bjørn Lomborg, recommend a climate policy approach that eschews mitigation altogether in favor of a response that relocates environmental refugees fleeing their damaged homes and deluged homelands, Cannavò defends the value of place as not so readily fungible as those observers suppose. People are not *placeless*, he argues, and our homes are more than replaceable commodities to be relocated whenever it is efficient to do so. No event in recent American environmental history more dramatically illustrates both the value that people attach to place and the perils that such commitments entail than the immediate and medium-term aftermath of hurricane Katrina. Focusing on New Orleans, which was once perhaps the most unique city in North America but that is now faced with rebuilding prospects that threaten to leave it much less racially, ethnically, and economically diverse, Cannavò suggests that there may be a "tragic dilemma" between the value of place and ecological responsibility. As climate change more generally is expected to do, the greatest costs of Katrina were borne by the poorest residents of New Orleans, he notes, and the city's environmental vulnerability was ironically increased by past efforts to improve nature, such as the reengineering of the Mississippi River Delta and the filling in of wetlands that served flood-control purposes. Lost in the wake of Katrina were not only built structures, he argues, but place-based communities and social networks, and the latter may be even more difficult to replace through "adaptation" efforts. As Cannavò notes, New Orleans "had to be engineered into existence," and now stands as "an embodiment of humanity's attempt to conquer nature," bringing a negative lesson to our appreciation of sustainability, because a stable home or homeland requires a "complex relationship with nature" that includes a combination of "elements of conquest and cooperation" with its forces. In post-Katrina New Orleans, he suggests, the value of place "collides" with that of environmental sustainability, and the two lead to opposite positions on the question of whether the city should ever have been built in the first place, and whether it should now be rebuilt. This dilemma can be lessened, if not avoided altogether, Cannavò argues, by increasing the policy role of mitigation that diminishes the need for adaptation.

Acknowledgments

This book would not have been possible without the patience and diligent efforts of its contributors or the help of Clay Morgan at The MIT Press. Thanks are also due to David Schlosberg for organizing the panels that served as the impetus for this theoretical discussion, to the two anonymous referees for their helpful comments, to White Horse Press and Taylor & Francis for allowing previously published work from *Environmental Values* and *Environmental Politics* to be included in this volume, and to Oxford University Press for allowing me to adapt sections from my *Atmospheric Justice* for use here. Thanks also to Gülay Uğur Göksel Yaşar for her assistance in manuscript preparation.

Notes

1. Elisabeth Rosenthal, "UN Report Describes Risks of Inaction on Climate Change," *New York Times*, November 17, 2007, A1.

2. Intergovernmental Panel on Climate Change (IPCC), *Climate Change 2007: Synthesis Report*, "Summary for Policymakers" (draft report), November 16, 2007, 20.

3. The 2006 Asian Pacific Partnership on Clean Development and Climate, which contains no binding emission caps or timetables, is widely regarded as an effort by Kyoto Protocol holdouts Australia (which, following the Labor Party's 2007 victory over the anti-Kyoto Liberal government, subsequently ratified the treaty) and the United States to undermine the UN framework convention process. U.S. Senator John McCain (R-AZ) described the pact as "nothing more than a nice little public relations ploy," and *The Economist* denounced it as "patent fig-leaf for the refusal of America and Australia to ratify Kyoto." See http://www.asiapacificpartnership.org/.

4. On the U.S. role in obstructing global climate policy efforts from the mid-1990s through 2007, see Steve Vanderheiden, *Atmospheric Justice: A Political Theory of Climate Change* (New York: Oxford University Press, 2008).

5. United Nations, *United Nations Framework Convention on Climate Change* (New York: United Nations 1992).

6. Intergovernmental Panel on Climate Change (IPCC), *Climate Change 2001: Synthesis Report*: Cambridge: Cambridge University Press, 2002), 12.

7. Chris Mooney, "Warmed Over," *The American Prospect*, online ed., January 10, 2005.

8. Vanderheiden, *Atmospheric Justice*.

9. Henry Shue, *Basic Rights* (Princeton, NJ: Princeton University Press, 1980).

Contributors

Martin J. Adamian
Department of Political Science
California State University,
Los Angeles

John Barry
Department of Politics, International
Studies, and Philosophy
Queens University Belfast

Peter F. Cannavò
Department of Government
Hamilton College

Stephen Gardiner
Department of Philosophy
University of Washington

George A. Gonzalez
Department of Political Science
University of Miami

Amy Lovecraft
Department of Political Science
University of Alaska Fairbanks

Timothy W. Luke
Department of Political Science
Virginia Polytechnic Institute and
State University

Leigh Raymond
Department of Political Science
Purdue University

Steve Vanderheiden
Department of Political Science
University of Colorado at Boulder

I

Justice, Ethics, and Global Climate Change

1

Allocating the Global Commons: Theory and Practice

Leigh Raymond

Many scholars and commentators agree that any future climate change treaty must be fair or equitable. This is both a practical and a moral imperative: moral because of the high economic and environmental stakes involved, and practical in the sense that any future treaty that is widely perceived as unfair is also unlikely to be ratified in the Hobbesian world of international relations. Of course, what constitutes a "fair" treaty is a point of greater controversy, one that has already sparked a wide-ranging debate over intertwined principles of ecological and distributive justice. The past two decades have witnessed a flowering of diverse normative arguments about climate "equity," many lovingly nurtured and carefully tended. Nor has this work been restricted to academics: policymakers and treaty negotiators have also considered the finer points of competing allocation schemes in great detail.

More often than not, the debate is cast as a question of principles by which we should distribute (or "allocate" in the language of the literature) the earth's finite capacity to absorb greenhouse gases, especially carbon dioxide. This atmospheric resource is sometimes referred to as part of the *global commons*—natural resources that remain beyond the control or ownership of any individual, corporation, or nation. Technically speaking, the phrase does some violence to the true idea of a *commons*, a term widely recognized as representing collectively owned property (by some limited community of individuals), rather than a resource beyond any ownership or control mechanism.[1] Many scholars, in fact, would probably refer to the global commons as a "global open-access resource" or "global nonproperty."[2] But for reasons of familiarity or linguistic appeal, the phrase "global commons" has stuck and will be used here as well.

Such extensive attention to normative questions of political theory is uncommon in the world of public policy scholarship, where considerations of efficiency and vested economic interests tend to rule the roost.[3]

All this should be of substantial encouragement to those seeking better integration of the ideas of normative theory and the study and practice of policymaking. Yet despite the many books, articles, white papers, and diplomatic "nonpapers" on the issue, progress toward a larger consensus on the qualities of an equitable climate change treaty remains elusive. The Kyoto Protocol staggered into legal effect in 2005, despite the non-participation of the largest contributor of greenhouse gases (GHGs) on the planet (on either a gross or per capita basis), the United States. While a heroic tale in its own right, the tremendously difficult process of ratifying the protocol (not to mention the significant question of whether signatories will actually meet their stated emission goals) gives one little reason to think that the right normative principles have been discovered to solve this political stalemate.

Indeed, some commentators despair of our ever finding such principles. David Victor, for instance, has called the allocation question the "Gordian knot" of climate change politics, and is pessimistic about the chances for any politically feasible solution.[4] Others are skeptical for different reasons, including a basic mistrust of any approach to solving the climate problem that smacks of the market and "enclosing" the global commons and "carbon colonialism" by creating private "rights to pollute."[5] Better to rely on other approaches to protecting natural resources outside the ownership or control of any individual or nation, these arguments maintain, than to give in to the siren song of "free market environmentalism" or its conceptual bedfellows.

Much of this pessimism seems well founded. Previous allocations of GHG emissions have been difficult, and the risk of abuse of market-based instruments is real.[6] Yet market-based approaches do offer some important advantages, potentially reducing compliance costs and easing the path toward difficult political agreements.[7] In this respect it is important to note that the world *has* dealt with difficult allocation issues in the past, relying on a variety of normative principles to guide distributions of private rights to portions of the global commons. Thus, this chapter attempts to bridge the normative and empirical worlds of political analysis by comparing the normative literature on climate change equity to these previous allocation experiences. The hope is that by making this connection, the chapter will use past experience to provide new insight into which normative arguments are more or less politically plausible in the climate change policy context. This in turn might spur new insights among normative thinkers about climate change equity, as they consider concrete examples of specific allocation ideas in action.

The discussion proceeds in four parts. First, it briefly reviews some of the more prominent normative arguments for allocating rights to emit GHGs. It then reviews a few previous allocations of portions of the global commons, including UN treaties regarding the deep ocean, Antarctica, and objects in outer space, looking for applied examples of the normative arguments mentioned in part I. Noting the relative scarcity of international allocations, the chapter then goes on to consider a wider range of normative principles in evidence in national allocations of GHG emissions, again with an eye on the normative principles outlined in part I. Finally, the chapter draws a few conclusions about which allocation principles seem more or less politically feasible given these past experiences, as well as raising a few important questions in this area for future research.

Prominent Allocation Arguments, Scholarly and Otherwise

There has been a remarkable outpouring of applied normative thinking regarding the question of allocating GHG emissions, dating back to the signing of the UN Framework Convention on Climate Change in 1992 and earlier. Although a full recounting of this literature is beyond the scope of this discussion,[8] it is possible to summarize much of this thinking as a handful of prominent "families of arguments" that have been made by multiple authors and parties to the treaty-negotiation process. Most of these arguments cite one form of equality or another as a guiding principle. But as Ronald Dworkin[9] and Amartya Sen[10] have noted, any such argument begs a basic question: "Equality of what"?

Keeping in mind, then, that the list is neither exhaustive nor evaluative, let us consider a few types of allocation proposals and their ethical underpinnings.

Equal Burdens
This principle is actually a combination of two prominent norms of entitlement: one based on a Humean notion of ownership via possession, the other on a more traditional Lockean idea of ownership from beneficial prior use. Both tend to be used to justify allocations based on recent emission levels—arguing that nations are entitled to their current levels of GHG emissions, or at least that current levels are the baseline from which any subsequent reductions must be prorated. They differ critically, however, in the nature of the normative argument offered. A Lockean claim of ownership argues that prior use of a resource adds value through

productive labor, thereby justifying a claim of "unilateral appropriation"—ownership without consent by other rivals. Thus, Lockean allocations based on prior use are seen as largely *prepolitical* or natural rights that are recognized, rather than created, by any political action allocating formal legal rights to a natural resource. Of course, Locke famously qualified this right of unilateral appropriation, including the requirement to leave "enough and as good" for others. Whether this "Lockean proviso" is met in any climate change allocation context is debatable.

More common in the climate change context is the argument for allocation based on possession rather than prior use. This line of thought echoes David Hume's view of ownership as an acceptable Pareto improvement for society. For Hume, property rights mean everyone gains by no longer fighting over resources on a daily basis, but the only way everyone would agree to such an arrangement is by ratifying the existing holdings of each actor. This is such a limited normative argument that some have hesitated to call it a "moral" argument for ownership at all.[11] Yet in many cases, allocation based on possession and the economic improvement that such security of ownership would entail is a prominent argument in the climate change policy realm. Several nations made essentially Humean arguments regarding GHG allocation in the negotiations leading up to the Kyoto Protocol,[12] and prominent arguments citing a need for equal "burden sharing" of GHG emission reduction among all nations rely implicitly or explicitly on this argument from possession.[13] Those most interested in the efficiency gains of clearly defined rights to emit GHGs, especially ones that can then be traded among nations, tend to fall into this category.

Equal Efficiency
A second family of allocation arguments focuses not on *total* GHG emissions but on the emission *rate*. Commentators and national leaders have both cited GHG emissions per unit of productive work as an allocation option. Common proposed measures include GHG emissions per unit of economic output (GDP) and per unit of energy produced (mBTUs). Arguments in this vein sometimes note that while the United States is one of the world's largest emitters of CO_2 on a per capita basis, its emissions per unit of GDP are actually quite average globally, and lower than some major developing nations.[14] More generally, this family of arguments includes the idea of *benchmarks*—environmentally acceptable emission rates. A benchmarked allocation might give each nation a fixed amount of GHG allowances per unit of economic production, for instance, regardless of that nation's historic emission patterns. This approach

rewards cleaner nations with low-emitting energy sources, while punishing those with dirtier, higher-carbon sources of economic activity (or at least providing a strong incentive for them to clean up their act). Such arguments were quite common in national negotiating position papers leading up to the Kyoto Protocol,[15] and are prevalent in the academic literature on climate equity.[16]

Equal Rights

A third major family of climate change allocation arguments relies on the principle of equal human rights. Under this perspective, each citizen of the world is entitled to an equal share of the atmosphere's ability to absorb GHGs as a matter of basic distributive justice. Most commonly, this argument is applied to the climate change case through a global allocation of emission rights to each nation proportional to its current or recent population. Under such a scheme, populous but relatively poor nations like China and India would become entitled to larger shares of the world's GHG emissions, while high per capita emitters like the United States would face a substantial reduction. Variants of the equal per capita argument offer different transition pathways from current emission patterns to an equal per capita future.[17] More radical versions seek an equal distribution per capita of emissions based on *historic* use of the atmosphere rather than current use. This "natural debt" argument points out that the long atmospheric lifespan of many GHGs means that historically high emitting nations have actually used an even greater share of this global sink capacity than current emissions indicate.[18] Closely related to the widely cited "polluter pays" idea, the natural debt view mandates even fewer emission rights to developed nations than a standard equal per capita argument.

Despite representing a strong and rather unprecedented egalitarian perspective in international relations, the idea of equal per capita shares of the atmosphere has actually received substantial attention in the academic literature, both positive[19] and critical.[20] Even more surprising is the fact that international negotiators appear to be taking the principle seriously, although it has yet to be reflected in any active climate change policy.[21]

Equal Subsistence Rights

A fourth family of allocation arguments is a variant of the equal rights idea. It also insists on an equal per capita distribution of rights to emit GHGs, but not *all* such rights. Most commonly, this idea is articulated through a distinction between subsistence and luxury emissions,[22] with subsistence emissions necessary for maintaining a basic standard of living to be distributed equally among nations according to population. How the remaining "luxury" emissions might be distributed is then open to further debate, with candidates including principles based on efficiency, possession, and prior use mentioned above. Drawing implicitly on a Kantian conception of property, the subsistence rights argument offers an appealing mix of ethical idealism and political realism.[23] It has also been the subject of significant criticisms, not least of which is the difficulty of distinguishing a "subsistence" emission from a luxury-based one.[24]

A prominent additional line of argument goes beyond these four allocation ideas. Some authors take a harder line on the allocation question, rejecting any distribution of private or national emission rights as an ethically bankrupt idea.[25] Skeptics of the idea condemn the creation of private rights in global, public resources as yet another effort by developed nations to appropriate natural resources without adequate consultation or compensation for developing nations.[26] Instead, this argument continues, global resources should remain outside exclusive private or national control, to be regulated in a manner beneficial to humanity as a whole.[27] Although this criticism of emission rights has weakened in domestic policy contexts of late, it remains prominent in discussions of international climate change treaties and responsibilities.

Although they cover a wide range of ideas, these five broad allocation arguments do not exhaust the subject. Other suggested allocation principles include ability to pay, vulnerability to future climate change impacts, and even quasi-Rawlsian principles maximizing the benefits to the least well-off in global society.[28] If current trends are any indication, even more ideas are likely in the future. But despite this plethora of dialogue and debate over how we should distribute GHG emissions, substantial skepticism remains over whether *any* such allocation is realistic or desirable. Given that the nations of the world have already dealt with global allocation problems for other resources, however, it seems worthwhile to review these experiences in pursuit of allocation strategies that might be both normatively defensible (at least according to some portion of this voluminous literature) and politically plausible.

International Allocations: Enclosing the Global Commons

Resources counted as part of the global commons have varied substantially over time. In 1982, for instance, the nations of the world extended national control over nearly a third of the earth's surface under the UN

Convention on the Law of the Sea (UNCLOS). After that dramatic action, one might say, the global commons was a little less "global." But it would be a mistake to assume that the process is exclusively one of growing privatization and national enclosure. Any regime of ownership or control includes significant costs, and those costs mean that if the resource loses sufficient value it may revert to open-access status. Many natural resources that were jealously guarded in their time are now effectively nonproperty once more.[29]

Nevertheless, at different times nations have struggled to allocate exclusive rights of ownership and control over portions of the global commons. Key examples include rights to the Antarctic continent, rights to ocean fishing and minerals, and rights to objects in outer space. In each instance, the allocation process led to different results based on different normative underpinnings. This section will consider each of these experiences in turn.

The Antarctic Treaty

During the early twentieth century, foreign exploration of Antarctica gave rise to a rash of territorial claims. In the 1920s and 1930s, Australia, France, New Zealand, Norway, and the United Kingdom all asserted sovereignty over "pie-slice" sectors of the continent, generally delineated by certain east-west longitudes extending from the Antarctic Circle to the South Pole. Two South American nations, Argentina and Chile, complicated things considerably in the early 1940s when each made claims overlapping the British sector and each other's.[30] The United States meanwhile considered making a similar claim on several occasions but decided against doing so. Instead, both the United States and the former Soviet Union opted to make no claims of their own or to recognize those of anyone else.[31]

National claims to portions of Antarctica relied on several normative principles. Possession and use of the continent were leading factors in establishing or publicly justifying such claims, as demonstrated by otherwise nonsensical actions regarding the placement and staffing of year-round scientific research stations, providing mail service to Antarctic bases, and moving spouses and children of scientists onto the continent.[32] An additional rule went beyond possession to invoke geographic proximity as a claim of entitlement. Thus, Southern Hemisphere states including Chile, Argentina, Australia, and New Zealand articulated claims to parts of Antarctica based on its relative proximity to their own national borders.[33]

The dispute over allocating Antarctica was postponed indefinitely, however, by the Antarctic Treaty of 1959. Momentum for the treaty began in 1957, with proclamation of the International Geophysical Year (IGY) including an informal agreement in Antarctica that scientists could move freely about the continent and build bases without regard to political complications.[34] Both politicians and scientists looked to preserve and formalize this peaceful coexistence at the end of the IGY, and to ease the ongoing international dispute over the continent. The result was the treaty, which ratified this informal principle of unrestricted access into international law.

The treaty is a brief document, guaranteeing access to the entire continent for scientists and banning all military activity. The most crucial element, however, is Article IV. This clause essentially pushes aside the formal allocation issue, stating that nothing in the treaty is a renunciation of past claims over parts of the continent, forms any basis for those claims, or prejudices the recognition or nonrecognition of such claims. Furthermore, no action taken while the treaty is in force can establish a further basis for making or denying a claim, and no new or expanded claims of sovereignty by signatories can be made.

Since that time, the "truce" over further definition or assertion of private claims to Antarctica has held. Indeed, the notion of mooting such claims permanently by making the continent a world park has gained some momentum, especially with the 1991 addition to the treaty of a protocol banning all mining activity on the continent.[35] Such an action would represent a rejection of any private or national rights, allowing Antarctica to remain a protected part of the global commons indefinitely. Despite such rhetoric, national claims to Antarctic resources remain an ongoing, if dormant, threat to any such arrangement. In this respect, Antarctica serves as a remarkable example of an international allocation process frozen in media res.

The UN Convention on the Law of the Sea (UNCLOS)

UNCLOS represents one of the most spectacular international allocations of a global resource. Finalized in 1982, after decades of negotiation, the convention was ratified by a sufficient number of nations to become legally effective in 1994.[36] A sweeping document, it addresses a number of issues including freedom of navigation, environmental protection, and issues of scientific research. In addition, it provides guidance on the allocation of marine and seabed resources among nations.

The treaty extends national sovereignty over the oceans in a dramatic manner. Under its terms, coastal nations are granted an Exclusive

Economic Zone (EEZ) extending 200 miles from their shorelines. Within this zone, all marine and seabed resources are reserved exclusively for the bordering nation. This is a distribution based on proximity much like the assertions made by Southern Hemisphere nations in the case of Antarctica. The EEZ provisions represent a breathtaking reduction of the global commons, granting control of an additional 35 or 36 percent of the planet's surface area to individual nations.[37]

Beyond this 200-mile boundary, however, the treaty allocates no additional national rights. The inability of sovereign nations or others to appropriate the "high seas" in this respect dates back several centuries, so the lack of exclusive rights to this resource is not a new idea.[38] In the case of fishing, the result is that marine life beyond an EEZ remains an open-access resource (or "nonproperty") lacking general regulatory authority. The floor of the deep ocean, however, is treated somewhat differently, being declared part of the "Common Heritage of Mankind (CHM)."[39] Any private rights to the minerals located there are seriously limited by the requirement that they generate benefits for "mankind as a whole," including both coastal and landlocked states.[40]

First suggested in 1967 by UN Ambassador Arvid Pardo of Malta, the CHM idea remains somewhat ambiguous in international law.[41] In general, it stands for a rejection of any private rights to the resource in question, supporting instead some (possibly unspecified) form of resource protection and development with benefits distributed to all nations regardless of priority, location, or other factors.[42] The CHM idea explicitly rejects any private or national appropriation of the resources in question. Thus, in the UNCLOS example, the treaty contemplates some sort of international body that would control all deep-sea mining, with whom any private or national entity must share all profits. These profits would then be distributed globally among all nations, coastal or not, by some formula yet to be determined.

Those opposing the CHM idea in the UNCLOS treaty sought greater recognition for those who actually developed the resources in question.[43] In essence, this view rejected the common-heritage approach for an open-access resource that might eventually be subject to private-ownership claims based on Lockean, prior-use principles. Opposition to the CHM idea was significant enough to prevent the United States and a handful of other industrialized states from signing or ratifying the treaty.[44] Despite a subsequent agreement effectively amending the convention by weakening the common-heritage approach to the deep-sea bed, the United States has yet to ratify.[45]

The Moon Treaty

A second example of the common-heritage idea in action was the 1979 *UN Agreement Governing the Activities of States on the Moon and Other Celestial Bodies*, more commonly known simply as the "Moon Treaty." The idea of the moon and other resources in outer space being part of the global commons dates back at least to the 1950s.[46] The 1967 Treaty on Activities in Space, now widely considered the basis of all international space law, included significant language preventing any private or national appropriation of extraterrestrial resources, and required their eventual development to be for the benefit of all humankind.[47] Like the high seas (and unlike the Antarctic case), there were no preexisting claims to impede the common-heritage idea in space. Nevertheless, as with the Law of the Sea treaty, the negotiations over the Moon Treaty were lengthy and difficult, particularly regarding the "common-heritage" phrase itself.[48] Giving up the possibility of future entitlements to these resources, presumably on the basis of future possession or prior use, remains a contentious issue among nations.[49]

Despite these difficulties, the final treaty that opened for signatures on December 18, 1979, included the CHM language. Future resource development on the moon could only occur under the guidance of an international authority, with an "equitable" sharing of the benefits among all nations. Reception to the Moon Treaty was frosty in the United States, and to date few nations have ratified the agreement, making its future relevance in international law uncertain.[50] Nevertheless, it remains a prominent parallel to the Law of the Sea in adopting the CHM alternative to private appropriation of these currently open-access resources.

Thus, international allocations of the global commons differ significantly from some of the ideas suggested in the literature on climate change. There is little evidence, for example, of distributions based on equal burdens or efficiency. Instead, two principles stand out as dominant:

1. Allocation based on possession and proximity
2. Rejection of any enclosure in the service of equal human rights to global resources

More remarkable still is the fact that allocation based on prior use, a dominant principle in many other distributive contexts, is of relatively little importance in these examples. Instead, it is the Humean argument of possession that faces off against a more radical, egalitarian rejection of any exclusive control over the earth's common resources that does not benefit all citizens of the world. Interestingly, a very similar conflict between possession and equal human rights to the atmosphere is at the heart of the current international debate over GHG emission allowances, as will be discussed further below.

GHG Allocation Rules: Enclosing the "National Commons"

While no treaty has fully allocated the atmospheric capacity to absorb GHGs at a global scale (the Kyoto allocation was among developed nations only, following the dictates of the 1995 "Berlin Mandate"), there have been important efforts at the national level. Most prominent are the National Allocation Plans (NAPs) undertaken by members of the European Community as part of the recently adopted EU Emissions Trading System (ETS). The EU ETS creates emission targets and allowances for GHGs in numerous European nations and allocates them on a national basis to pollution sources at the installation level. The ETS then allows companies to trade allowances throughout the European Union in order to lower the overall cost of meeting the community's collective Kyoto obligation to lower CO_2 emissions by 8 percent from 1990 levels.

Brought online in 2005, the EU ETS is the largest experiment in GHG emission trading to date and the first major opportunity to allocate GHG emission rights at a national level. Interestingly, the most common allocation principles in the ETS are quite different from those discussed in the international examples presented above. Rather than equal human rights, or rules based on possession or prior use, the EU NAPs tended to rely on projections of economic need, combined in limited cases with auctions and benchmarks of desirable environmental behavior. A brief discussion of each of these principles follows, focusing on the United Kingdom as an instructive example. Although other NAPs varied to some degree from the UK plan, the basic principles were relatively similar across the entire European Union.[51]

Economic Need

The UK NAP took a two-stage approach to allocating emission allowances. First, it assigned allowances to entire sectors of the economy (e.g., the pulp and paper industry). Then, in a second set of decisions, the plan assigned allowances to specific installations within each economic sector. As it divided the pie in this manner, the government relied on different principles for different stages of the process. At the sectoral level, the UK government relied on complex economic projections of each industry's likely future needs. Thus, some industries received larger relative portions of the total allowance pie based on projections of increasing economic activity, while others received smaller allocations based on negative trends. The energy sector was then forced to meet any remaining reductions from the status quo required of the United Kingdom under the EU Burden Sharing agreement, based in large part on the sector's relative insulation from international competition.

By contrast, installation-level allocations were based primarily on recent historic emission levels, prorated if necessary to meet the new sector limit. In this manner the UK NAP allocated at finer scales based on current possession or use, rather than calculations of future need (although the two were clearly expected to be connected). Indeed, need also remained a direct part of the allocation formula at the installation level: installations that ceased operations after the NAP went into effect were required to surrender their future allowances, rather than retaining them in future years as an asset to be sold or transferred to other operators.

Environmental Efficiency

A second idea that was considered seriously for the UK NAP, but largely rejected, was an allocation based on benchmarks. As noted above, the typical justification for benchmarking is that government should not reward "dirty" companies or industries for bad behavior, as allocations based on previous emission levels or unqualified estimates of future need risk doing. Instead, a benchmarked allocation can level the playing field by granting allowances to all facilities based on a constant emission rate multiplied by the facility's historic record of energy production or consumption.

Some UK industries and sectors favored benchmarking as a fairer way to allocate emissions. In the end, however, benchmarks played only a small role in the final NAP, used for allocations to new future sources in the market, but having little impact on the distribution of allowances to current emitters. While other EU NAPs invested a greater effort in benchmarking allocations, in the end many of them abandoned the effort as too difficult to implement on a larger scale, at least for the time being. According to a recent survey of all EU NAPs, the difficulty of determining a fair and acceptable benchmark for a wide range of industries and economic sectors was a major reason why this allocation argument failed to be used more widely.[52] Interestingly, this problem illustrates how arguments based on benchmarks can resemble those based on equal subsistence emissions, at least in terms of posing similar definitional challenges regarding an appropriate emission standard to apply across the board. This surprising parallel will be discussed further below.

Auctions

A third idea relevant to the UK NAP was that of an auction: selling emission allowances to industries up to the limit imposed by an annual national emission "cap." According to the EU Commission on the Environment, no NAP could auction more than 5 percent of the total allowance distributed in the first "phase" of the program (from 2005 to 2008). That limit increased to 10 percent for Phase 2 NAPs, which were due to be completed in 2007. Despite some discussion of auctioning as an allocation option, the United Kingdom ultimately declined to auction any allowances in the first phase of the program.

Auctions are generally the favored allocation strategy of economists considering an emissions trading program. They provide an efficient distribution of emission credits at the outset, and provide strong financial incentives for industries to be creative and aggressive in reducing their emissions. Yet auctions remain unpopular politically, for reasons both obvious and subtle. At the simplest level, regulated industries dislike new taxes imposed on their operations, which is what an auction of allowances amounts to. Emissions trading programs are much easier to promote politically when the affected industries receive their allowances free of charge. Yet not every public resource is given away by the government to private actors. Recent auctions of broadcast-spectrum frequencies have raised hundreds of millions of dollars of revenue for governments in the United States and European Union. Sometimes, governments are able to auction off portions of the global or national commons without difficulty, whereas in other instances such sales are nigh on impossible.

A key factor in this distinction seems to go back to the idea of possession and prior use. Auctions appear to be more politically palatable when industries are not currently using the resource in question; charging companies to use a resource that they have already been exploiting free of charge is a more challenging proposition. Given this general pattern (which clearly does not hold across every example—there being plenty of government giveaways of "unused" resources as well), and the initial experience of the UK NAP, one might be pessimistic about counting on auctions as a viable allocation strategy for GHG emission rights in the future. And indeed, there has been precious little discussion of auctions in the UNFCCC negotiations to date.

And yet there are signs of change on the horizon that might give would-be auctioneers some hope. Following up on earlier government rhetoric, the UK is planning to auction 7% of all emission allowances in Phase 2 of the process, for the years 2008–2012.[53] That change represents a relatively rare instance of government charging an industry for the right to use a resource previously enjoyed free of charge. If that trend continues, auctions will become an increasingly plausible allocation option.

Implications and Future Directions

Several important points seem to emerge from the preceding juxtaposition of actual and theoretical models of allocation for various resources in various political contexts. The first, and most basic, is that the nations of the world have managed to make headway on global allocation problems in the past, which should qualify some of the existing pessimism regarding the ability to do so in the climate change context. It is true that allocating atmospheric emission rights is different from allocating (say) fishing rights: GHG emissions are more intangible, less easily linked to existing ideas of sovereignty and ownership, and more economically critical. Nevertheless, the partial and incomplete allocations of the global commons described in this chapter indicate that at least some of the time, the nations of the world can reach agreement on the distribution of national rights to formerly open-access resources.[54]

More surprising and significant, perhaps, is the limited role of prior resource use in the allocations reviewed here. Claims of entitlement based on prior use have been central to many allocations of natural resources, particularly in the United States, where the idea of entitlement based on "beneficial use" has underwritten public laws distributing land, water, range forage, timber, and hardrock minerals, among others. Yet in the examples discussed in this chapter, Lockean allocations based on prior use are surprisingly scarce. In the international cases, allocations largely gravitated toward Humean principles based on possession and proximity or an egalitarian principle demanding a share of global resources for everyone based on their "common humanity." At the national scale, the United Kingdom distributed GHG emissions first and foremost based on estimations of *future* need rather than *prior* use of the resource. Even in instances where there was extensive prior use by competitors, like foreign fishing in EEZs off the coast of a given country, negotiators declined to recognize that use as relevant or legitimate. This represents a significant pattern of deviation from the Lockean ideal.

The absence of Lockean allocations suggests that a window of opportunity may be opening for alternative distribution strategies, even for resources with a history of prior use. Instead, three alternative allocation principles seem to be gaining credibility: economic need, environmental efficiency (benchmarks), and willingness to pay (auctions). Rules based on benchmarks were a key part of the allocation of SO_2 emission allowances in the United States under the 1990 acid rain program, and were a popular idea for discussion in creating the first set of NAPs for the EU ETS. Given that high level of interest, it seems likely that future EU NAPs will find a way to take advantage of benchmarked standards as part of their allocation strategies. Auctions, too, seem likely to play a growing role in the process, as governments slowly shift toward different property-based norms that define the atmosphere as a public resource rather than something already owned by private industries and companies.

Even more striking, however, is the dominant role of economic projections in the UK NAP. The United Kingdom's privileging of need as an allocation principle supports a very different allocation argument in a surprising manner: the demand for subsistence emissions for every *person* on the planet based on their future economic needs. While the recipients are quite different, the basic normative argument is parallel: GHG allowances are something industries and individuals are entitled to based not on their previous actions or current qualities, but rather on their future requirements for success. Thus, the UK NAP (and other EU NAPs following a similar methodology) actually opens the door for more serious consideration, one could argue, of need-based arguments in general, including those supporting global allocations of subsistence emissions for all.

This odd pairing brings to mind another surprising conceptual link: the relationship between equal per capita allocation arguments and treaty language embracing the Common Heritage of Mankind principle. Recall that the CHM idea rejects any private allocation of the global commons, preserving its public status for the benefit of all nations. This principle of nonallocation, if you will, is more immediately evocative of arguments by critics of market-based policies in general than it is of arguments for any particular distribution of those resources, like the equal per capita position.

Yet the deeper links between the CHM and equal per capita arguments are worth noting. Both invoke a basic rejection of private, exclusive rights to natural resources that fail to benefit everyone in an egalitarian manner. In this sense both are strongly evocative of the property ideas of French political theorist P. J. Proudhon,[55] who argued that property rights and private ownership can only be justified based on universal consent, and that the only distribution to which such consent is imaginable is one of equal shares for all. Proudhon made this argument primarily as a critique of the entire institution of private property (which he famously derided as "theft"), but in so doing he blurred the line between a strictly egalitarian distribution of private rights and no distribution at all. In a sense, Proudhon helps us to see that two distinct normative positions—"property for all" and "no property at all"—are actually quite closely related.

Proudhon's insight is important to the modern debate over allocating climate change emissions. Currently, two distinct egalitarian arguments are prominent in the case of climate change, one espousing some sort of equal per capita distribution, the other rejecting any private rights in the resource at all. The CHM idea helps us to see that those two positions are more connected than they first appear, and indeed might be considered in some ways as two sides of the same coin. While of course there are other reasons why some object to the enclosure of the atmosphere in this manner, it seems important to recognize that not all enclosures are the same, and that some offer substantial egalitarian benefits that may satisfy objections to a market-based approach. Even Ronald Coase,[56] one of the intellectual godfathers of market-based policies, carefully noted that any initial distribution of rights would be efficient as long as the transaction and information costs were low. Critics of emissions trading in general might therefore consider Proudhon's argument and if a strictly egalitarian allocation might actually address some or most of their concerns.

In light of this discussion, where might empirical research on normative ideas about allocation go from here? The simplest answer is to continue exploring the relationship between political context, resource qualities, and the viability of various allocation principles. Thus, a simple but relevant question is asking why auctions are politically feasible in some contexts but unthinkable in others. Why, in other words, are nations able to auction off millions of dollars of resources in some cases (e.g., broadcasting frequencies), but not charge a dime for private rights to public resources in others (e.g., emission allowances)? Some likely reasons already present themselves, like prior use being a strong disincentive to auctioning, but those explanations remain contingent and case-specific. It would be both fascinating and useful to see empirical work toward a broader theory of when specific allocation rules like auctions are more or less politically viable.

The cases discussed here also raise important questions of scale. Allocations vary significantly between the national and the international level—rules found in the EU ETS system are quite different from those followed at the level of the global commons. Yet the examples are conceptually related; indeed, the EU ETS system itself represents a multistate allocation process, with shared rules like the limit on auctions agreed to by representatives of multiple sovereign nations. Recoiling from the difficulties of the Kyoto process, some have already urged a new approach to climate change negotiations that builds on a "deep" commitment by a small group of dedicated nations, slowly expanding participation to eventually reach a "broad" coverage of the world's emissions.[57] Any such approach will have to consider how to expand allocation rules from a few nations to a much larger, international group of actors. Thus, exploring how and to what degree the set of plausible allocation rules seems to change as one moves from national to bilateral and multilateral contexts would be an important research goal.

At the same time, the idea of *individual* allocation rights is gaining attention from environmental policy entrepreneurs[58] as well as public officials.[59] As we get closer to a day when each person may own personal, transferable carbon credits, we will have to confront the difficult issue of allocation at the individual level. Here, research on the potential for scaling down national and international allocation principles to individuals will be important, much like the already-noted parallel between allocations to economic industries and individuals based on need. If equal per capita shares are distributed somewhere at an individual level, can international agreements on the same principle be far behind? Or does the set of credible allocation rules change significantly from one scale to another?

Finally, future interactions between allocation rules and technological developments bear further consideration. Traditionally, cap and trade policies or other market-based approaches are designed to create incentives for new pollution-control technologies, without circumscribing what those technologies might look like. Thus, the U.S. acid rain program lowered the price of scrubbing SO_2, even as it also spurred development of new technologies to allow the burning of low-sulfur coal in power-plant boilers. This is familiar stuff to students of environmental policy, but a less familiar question is this: What happens to the political viability of various market-based policies as the technologies for reducing or limiting emissions change over time? Will an allocation based largely on projected economic need, for instance, still be politically viable if cheap and reliable carbon sequestration technology comes online? How will potential advances in generating carbon-free energy affect dominant norms of entitlement to the global carbon sink? Do the same kinds of allocation rules seem likely to apply to allocations of carbon *storage* credits (in terrestrial biomass like soil or forests, for instance) as for emissions? New policies and new technologies dealing with these issues are on the horizon. Those seeking to understand the allocation process will therefore face a wide variety of technological and resource conditions in the near future that may or may not significantly affect which normative allocation arguments are conceptually and politically plausible. More work in this area seems vital.

Conclusion

The preceding discussion indicates that the process of natural resource allocation, while difficult, has been and continues to be addressed in a variety of ways by the nations of the world. The challenge of GHG emission allocation is an especially difficult one for any national or international policy dealing with climate change. Yet the allocation challenge also presents unique opportunities for promoting the integration of ecological and ethical interests, even within a market-based framework. Perhaps surprisingly, the nations of the world have shown sustained interest in different normative options for meeting that challenge. At the same time, certain allocation options have been dominant in some contexts, and beyond consideration in others. Clearly, we need to better understand the relationship between an allocation's political and economic setting and its ultimate outcome in order to better evaluate market-based options to environmental problems in general.

What stands out the most from this discussion, however, is the rather startling degree of conceptual creativity on display in addressing the issue of allocation, *both at an empirical and at a theoretical level.* Academics, advocates, and policymakers alike continue to generate fascinating and creative ideas for allocating emission rights to GHGs, as part of their larger efforts to address the problem of anthropogenic climate change. Rarely, it seems, has normative political theory had such an explicit and important role to play in the day-to-day practice of environmental policymaking. In this respect, the contributions to this book represent a rare synergy in political science between the theory and the practice of politics. Critical or sympathetic, theorists have a real chance to make an impact on policymaking with regard to distributing climate change rights and burdens. Similarly, they will benefit from paying close attention to various new and ongoing policy experiments struggling with those very issues. The potential for learning, on both sides, is exciting, uncommon, and vital to any lasting solution of one of the most serious environmental problems facing the world today.

Notes

1. E. Ostrom, *Governing the Commons: The Evolution of Institutions for Collective Action* (New York: Cambridge University Press, 1990).

2. D. W. Bromley, "Comment: Testing for Common versus Private Property," *Journal of Environmental Economics and Management* 21 (1991): 92–96; D. H. Cole, *Pollution and Property* (New York: Cambridge University Press, 2002).

3. M. Sagoff, *The Economy of the Earth* (New York: Cambridge University Press, 1988).

4. D. G. Victor, *Climate Change: Debating America's Policy Options* (New York: Council on Foreign Relations, 2004); D. G. Victor, *The Collapse of the Kyoto Protocol and the Struggle to Slow Global Warming* (Princeton, NJ: Princeton University Press, 2001).

5. T. W. Luke, "The System of Sustainable Degradation," *Capitalism, Nature, Socialism* 17, no. 1 (2006): 99–112; H. Bachram, "Climate Fraud and Carbon Colonialism: The New Trade in Greenhouse Gases," *Capitalism, Nature, Socialism* 15 (2004): 5–20.

6. R. Repetto, "The Clean Development Mechanism: Institutional Breakthrough or Institutional Nightmare?," *Policy Sciences* 34 (2001): 303–327.

7. D. Burtraw and K. Palmer, *The Paparazzi Take a Look at a Living Legend: The SO_2 Cap-and-Trade Program for Power Plants in the United States* (Washington, DC: Resources for the Future, 2003); R. E. Cohen, *Washington at Work: Back Rooms and Clean Air* (Boston: Allyn and Bacon, 1995).

8. For a more detailed summary, see S. M. Gardiner, "Ethics and Global Climate Change," *Ethics* 114 (2004): 555–600.

9. R. Dworkin, "What Is Equality? Part 1: Equality of Welfare," *Philosophy and Public Affairs* 10, no. 3 (1981): 185–251.

10. A. Sen, *Inequality Reexamined* (Cambridge, MA: Harvard University Press, 1992).

11. J. Waldron, "The Advantages and Difficulties of the Humean Theory of Property," *Social Philosophy and Policy* 11 (1994): 85–123.

12. L. Raymond, *Private Rights in Public Resources: Equity and Property Allocation in Market-Based Environmental Policy* (Washington, DC: Resources for the Future Press, 2003).

13. See the examples in Gardiner, "Ethics and Global Climate Change," or D. A. Brown, *American Heat: Ethical Problems with the United States' Response to Global Warming* (Lanham, MD: Rowman & Littlefield, 2002).

14. Victor, *Climate Change.*

15. Raymond, *Private Rights in Public Resources.*

16. Victor, *Climate Change*; H. E. Ott and W. Sachs, *Ethical Aspects of Emissions Trading* (Wuppertal, Germany: Wuppertal Institute, 2000); A. Rose and B. Stevens, "The Efficiency and Equity of Marketable Permits for CO_2 Emissions," *Resource and Energy Economics* 15 (1993): 117–146.

17. For example, A. Meyer, "The Kyoto Protocol and the Emergence of "Contraction and Convergence" as a Framework for an International Political Solution to Greenhouse Gas Emissions Abatement," in O. Hohmayer and K. Rennings, eds., *Man-Made Climate Change—Economic Aspects and Policy Options* (Mannheim: Zentrum für Europäische Wirtschaftsforschung (ZEW), 1999).

18. P. Singer, *One World: The Ethics of Globalization* (New Haven, CT: Yale University Press, 2002); K. R. Smith, "The Natural Debt: North and South," in T. W. Giambelluca and A. Henderson-Sellers, eds., *Climate Change: Developing Southern Hemisphere Perspectives* (Chichester, UK: Wiley, 1996).

19. T. Athanasiou and P. Baer, *Dead Heat: Global Justice and Global Warming* (New York: Seven Stories Press, 2002); P. Barnes, *Who Owns the Sky?* (Washington, DC: Island Press, 2001); A. D. Sagar, "Wealth, Responsibility, and Equity: Exploring an Allocation Framework for Global GHG Emissions," *Climatic Change* 45 (2000): 511–527; A. Agarwal and S. Narain, *Global Warming in an Unequal World: A Case of Environmental Colonialism* (New Delhi: Center for Science and Environment, 1991).

20. Ott and Sachs, *Ethical Aspects of Emissions Trading*; E. Claussen and L. McNeilly, *The Complex Elements of Global Fairness* (Washington, DC: Pew Center on Global Climate Change, 1998).

21. L. Raymond, "Viewpoint: Cutting the 'Gordian Knot' in Climate Change Policy," *Energy Policy* 34 (2006): 655–658.

22. H. Shue, "Subsistence Emissions and Luxury Emissions," *Law and Policy* 15, no. 1 (1993): 39–59; see also Ott and Sachs, *Ethical Aspects of Emissions Trading.*

23. Raymond, "Viewpoint: Cutting the 'Gordian Knot' in Climate Change Policy."

24. Gardiner, "Ethics and Global Climate Change."

25. Luke, "The System of Sustainable Degradation."

26. Bachram, "Climate Fraud and Carbon Colonialism."

27. B. Tokar, *Earth for Sale: Reclaiming Ecology in the Age of Corporate Greenwash* (Boston: South End Press, 1997).

28. Rose and Stevens, "The Efficiency and Equity of Marketable Permits for CO_2 Emissions."

29. Cole, *Pollution and Property.*

30. E. S. Milenky and S. I. Schwab, "Latin America and Antarctica," *Current History* 82 (1983): 52.

31. J. E. Mielke, *Polar Research: U.S. Policy and Interests* (Washington, DC: Congressional Research Service, 1996).

32. M. Parfit, "The Last Continent," *Smithsonian* 15 (1984): 50–60; F. M. Auburn, *Antarctic Law and Politics* (Bloomington: Indiana University Press, 1982).

33. Auburn, *Antarctic Law and Politics.*

34. P. J. Beck, *The International Politics of Antarctica* (London: Croom and Helm, 1986).

35. G. Porter and J. W. Brown, *Global Environmental Politics* (Boulder, CO: Westview Press, 1996).

36. M. A. Browne, *The Law of the Sea Convention and U.S. Policy* (Washington, DC: Congressional Research Service, 2000).

37. E. L. Miles, *Global Ocean Politics: The Decision Process at the Third United Nations Conference on the Law of the Sea 1973–1982* (The Hague: Martinus Nijhoff, 1998).

38. R. Hannesson, *The Privatization of the Oceans* (Cambridge, MA: MIT Press, 2004).

39. United Nations Convention on the Law of the Sea (UNCLOS): XI 136.

40. UNCLOS: IX 140.

41. K. Baslar, *The Concept of the Common Heritage of Mankind in International Law* (The Hague: Martinus Nijhoff, 1998).

42. Browne, *The Law of the Sea Convention and U.S. Policy.*

43. J. E. Mielke, *Deep Seabed Mining: U.S. Interests and the U.N. Convention on the Law of the Sea* (Washington, DC: Congressional Research Service, 1995).

44. G. V. Galdorisi and K. R. Vienna, *Beyond the Law of the Sea: New Directions for U.S. Oceans Policy* (Westport, CT: Praeger Press, 1997).

45. Browne, *The Law of the Sea Convention and U.S. Policy.*

46. K. Pritzsche, *Development of the Concept of "Common Heritage of Mankind," in Outer Space Law and Its Contents in the 1979 Moon Treaty* (Berkeley: School of Law, University of California at Berkeley, 1984).

47. K.-U. Schrogl, "Legal Aspects Related to the Application of the Principle That the Exploration and Utilization of Outer Space Should Be Carried Out for the Benefits and in the Interest of All States Taking into Account the Needs of Developing Countries," in M. Benko and K.-U. Schrogl, eds., *International Space Law in the Making* (Gif-sur-Yvette Cedex: Editions Frontières, 1993).

48. U.S. Senate Committee on Commerce, Science, and Transportation, *Report on Agreement Governing the Activities of States on the Moon and Other Celestial Bodies* (Washington, DC: Government Printing Office, 1980).

49. In this regard, it is worth recalling that one of the first acts of the Apollo astronauts on setting foot on the moon's surface was to plant a sovereign, American flag.

50. Pritzsche, *Development of the Concept of "Common Heritage of Mankind" in Outer Space Law and Its Contents in the 1979 Moon Treaty.*

51. A. D. Ellerman, B. K. Buchner, C. Carraro, eds. *Rights, Rents, and Fairness: Allocation in the European Emissions Trading Scheme* (New York: Cambridge University Press, 2007).

52. Ellerman, Buchner, and Carraro, *Rights, Rents, and Fairness: Allocation in the European Emissions Trading Scheme.*

53. L. Raymond, "Allocating Greenhouse Gas Emissions under the EU ETS: The UK Experience," paper presented at the Sixth Open Meeting of the Human Dimensions of Global Environmental Change Research Community, University of Bonn, Germany. 2005.

54. For more on this surprising outcome in the UNCLOS case, see also R. Hannesson, *The Privatization of the Oceans.*

55. P.-J. Proudhon, *What Is Property?* (New York: Cambridge University Press, [1840] 1993).

56. R. Coase, "The Problem of Social Cost," *Journal of Law and Economics* 3, no. 1 (1960): 1–44.

57. Victor, *Climate Change: Debating America's Policy Options.*

58. Barnes, *Who Owns the Sky?*

59. C. Clover, "Miliband Backs Idea of Carbon Rationing for All," *Daily Telegraph*, London, July 21, 2006.

2

A Perfect Moral Storm: Climate Change, Intergenerational Ethics, and the Problem of Corruption

Stephen Gardiner

There's a quiet clamor for hypocrisy and deception; and pragmatic politicians respond with . . . schemes that seem to promise something for nothing. Please, spare us the truth.[1]

The most authoritative scientific report on climate change begins by saying: "Natural, technical, and social sciences can provide essential information and evidence needed for decisions on what constitutes 'dangerous anthropogenic interference with the climate system.' At the same time, *such decisions are value judgments.*"[2] There are good grounds for this statement. Climate change is a complex problem raising issues across and between a large number of disciplines, including the physical and life sciences, political science, economics, and psychology, to name just a few. But without wishing for a moment to marginalize the contributions of these disciplines, ethics does seem to play a fundamental role.

Why so? At the most general level, the reason is that we cannot get very far in discussing why climate change is a problem without invoking ethical considerations. If we do not think that our own actions are open to moral assessment, or that various interests (our own, those of our kin and country, those of distant people, future people, animals, and nature) matter, then it is hard to see why climate change (or much else) poses a problem. But once we see this, then we appear to need some account of moral responsibility, morally important interests, and what to do about both. And this puts us squarely in the domain of ethics.

At a more practical level, ethical questions are fundamental to the main policy decisions that must be made, such as where to set a global ceiling for greenhouse gas emissions, and how to distribute the emissions allowed by such a ceiling. For example, where the global ceiling is set depends on how the interests of the current generation are weighed against those of future generations; and how emissions are distributed under the global gap depends in part on various beliefs about the appropriate role of energy consumption in people's lives, the importance of historical responsibility for the problem, and the current needs and future aspirations of particular societies.

The relevance of ethics to substantive climate policy thus seems clear. But this is not the topic that I wish to take up here.[3] Instead, I want to discuss a further, and to some extent more basic, way in which ethical reflection sheds light on our present predicament. This has nothing much to do with the substance of a defensible climate regime; instead, it concerns the process of making climate policy.

My thesis is this. The peculiar features of the climate change problem pose substantial obstacles to our ability to make the hard choices necessary to address it. Climate change is a perfect moral storm. One consequence of this is that, even if the difficult ethical questions could be answered, we might still find it difficult to act. For the storm makes us extremely vulnerable to moral corruption.[4]

Let us say that a perfect storm is an event constituted by an unusual convergence of independently harmful factors where this convergence is likely to result in substantial, and possibly catastrophic, negative outcomes. The phrase "perfect storm" seems to have become prominent in popular culture through Sebastian Junger's book of that name and the associated Hollywood film.[5] Junger's tale is based on the true story of the *Andrea Gail*, a fishing vessel caught at sea during a convergence of three particularly bad storms.[6] The sense of the analogy is then that climate change appears to be a perfect moral storm because it involves the convergence of a number of factors that threaten our ability to behave ethically.

Because climate change is a complex phenomenon, I cannot hope to identify all the ways its features cause problems for ethical behavior. Instead, I will identify three especially salient problems—analogous to the three storms that hit the *Andrea Gail*—that converge in the climate change case. These three "storms" arise in the global, intergenerational, and theoretical dimensions, and I will argue that their interaction helps to exacerbate and obscure a lurking problem of moral corruption that may be of greater practical importance than any of them.

The Global Storm

The first two storms arise out of important characteristics of the climate change problem. I label these characteristics

- Dispersion of causes and effects
- Fragmentation of agency
- Institutional inadequacy

Because these characteristics manifest themselves in especially salient dimensions—the spatial and the temporal—it is useful to distinguish two distinct but mutually reinforcing components of the climate change problem. I call the first the "global storm." This corresponds to the dominant understanding of the climate change problem, and it emerges from a predominantly spatial interpretation of the three characteristics.

Let us begin with the *dispersion of causes and effects*. Climate change is a truly global phenomenon. Emissions of greenhouse gases from any location on the earth's surface are fully dispersed through the atmosphere and then play a role in affecting climate globally. Hence, the impact of any particular emission of greenhouse gases is not realized solely at its source, either individual or geographic; rather, impacts are dispersed to other actors and regions of the earth. Such spatial dispersion has been widely discussed.

The second characteristic is the *fragmentation of agency*. Climate change is not caused by a single agent, but by a vast number of individuals and institutions not unified by a comprehensive structure of agency. This is important because it poses a challenge to humanity's ability to respond.

In the spatial dimension, this feature is usually understood as arising out of the shape of the current international system, as constituted by states. Then the problem is that, given that there is not only no world government but also no less centralized system of global governance (or at least no effective one), it is very difficult to coordinate an effective response to global climate change.[7]

This general argument is generally given more bite through the invocation of a certain familiar theoretical model.[8] For the international situation is usually understood in game-theoretic terms as a prisoner's dilemma, or what Garrett Hardin calls a "tragedy of the commons."[9] For the sake of ease of exposition, let us describe the prisoner's dilemma scenario in terms of a paradigm case, that of overpollution.[10] Suppose that a number of distinct agents are trying to decide whether to engage in a polluting activity, and that their situation is characterized by the following two claims:

PD1. It is *collectively rational* to cooperate and restrict overall pollution: each agent prefers the outcome produced by everyone restricting their individual pollution over the outcome produced by no one doing so.

PD2. It is *individually rational* not to restrict one's own pollution: when each agent has the power to decide whether they will restrict their pollution, each (rationally) prefers not to do so, whatever the others do.

Agents in such a situation find themselves in a paradoxical position. On the one hand, given PD1, they understand that it would be better for everyone if every agent cooperated, but, on the other hand, given PD2, they also know that they should all choose to defect. This is paradoxical because it implies that if individual agents act rationally in terms of their own interests, they collectively undermine those interests.[11]

A tragedy of the commons is essentially a prisoner's dilemma involving a common resource. This has become the standard analytic model for understanding regional and global environmental problems in general, and climate change is no exception. Typically, the reasoning goes as follows. Imagine climate change as an international problem and conceive of the relevant parties as individual countries, who represent the interests of their countries in perpetuity. Then, PD1 and PD2 appear to hold. On the one hand, no one wants serious climate change. Hence, each country prefers the outcome produced by everyone restricting their individual emissions over the outcome produced by no one doing so, and so it is collectively rational to cooperate and restrict global emissions. But, on the other hand, each country prefers to free ride on the actions of others. Hence, when each country has the power to decide whether it will restrict its emissions, each prefers not to do so, whatever the others do.

From this perspective, it appears that climate change is a normal tragedy of the commons. Still, there is a sense in which this turns out to be encouraging news, for, in the real world, commons problems are often resolvable under certain circumstances, and climate change seems to fill these desiderata.[12] In particular, it is widely said that parties facing a commons problem can resolve it if they benefit from a wider context of interaction, and this appears to be the case with climate change, since countries interact with each other on a number of broader issues, such as trade and security.

This brings us to the third characteristic of the climate change problem, *institutional inadequacy*. There is wide agreement that the appropriate means for resolving commons problems under the favorable conditions just mentioned is for the parties to agree to change the existing incentive structure through the introduction of a system of enforceable sanctions. (Hardin calls this "mutual coercion, mutually agreed upon.") This transforms the decision situation by foreclosing the option of free riding, so that the collectively rational action also becomes individually rational. Theoretically, then, matters seem simple, but in practice things are different. For the need for enforceable sanctions poses a challenge at the global level because of the limits of our current, largely national, institutions, and the lack of an effective system of global governance. In essence, addressing climate change appears to require global regulation of greenhouse gas emissions, where this includes establishing a reliable enforcement mechanism; but the current global system—or lack of it—makes this difficult, if not impossible.

The implication of this familiar analysis, then, is that the main thing needed to solve the global warming problem is an effective system of global governance (at least for this issue). And there is a sense in which this is still good news. For, in principle at least, it should be possible to motivate countries to establish such a regime, since they ought to recognize that it is in their best interest to eliminate the possibility of free riding and so make genuine cooperation the rational strategy at the individual as well as collective level.

Unfortunately, however, this is not the end of the story. For there are other features of the climate change case that make the necessary global agreement more difficult, and so exacerbate the basic global storm.[13] Prominent among these is scientific uncertainty about the precise magnitude and distribution of effects, particularly at the national level.[14] One reason for this is that the lack of trustworthy data about the costs and benefits of climate change at the national level casts doubt on the truth of PD1. Perhaps, some nations wonder, we might be better off with climate change than without it. More importantly, some countries might wonder whether they will at least be relatively better off than other countries, and so might get away with paying less to avoid the associated costs.[15] Such factors complicate the game theoretic situation, and so make agreement more difficult.

In other contexts, the problem of scientific uncertainty might not be so serious. But a second characteristic of the climate change problem exacerbates matters in this setting. The source of climate change is located deep in the infrastructure of current human civilizations; hence, attempts to combat it may have substantial ramifications for human social life. Climate change is caused by human emissions of greenhouse gases, primarily carbon dioxide. Such emissions are brought about by the burning of fossil fuels for energy. But it is this energy that supports existing economies. Hence, given that halting climate change will require deep cuts in projected global emissions over time, we can expect that such action will have profound effects on the basic economic organization of the developed countries and on the aspirations of the developing countries.

This has several salient implications. For one thing, it suggests that those with vested interests in the continuation of the current system—for example, many of those with substantial political and economic power—will resist such action. For another, unless ready substitutes are found, real mitigation can be expected to have profound impacts on how humans live and how human societies evolve. Hence, action on climate change is likely to raise serious, and perhaps uncomfortable, questions about who we are and what we want to be. Third, this suggests a status quo bias in the face of uncertainty. Contemplating change is often uncomfortable; contemplating basic change may be unnerving, even distressing. Since the social ramifications of action appear to be large, perspicuous, and concrete, but those of inaction appear uncertain, elusive, and indeterminate, it is easy to see why uncertainty might exacerbate social inertia.[16]

The third feature of the climate change problem that exacerbates the basic global storm is that of skewed vulnerabilities. The climate change problem interacts in some unfortunate ways with the present global power structure. For one thing, the responsibility for historical and current emissions lies predominantly with the richer, more powerful nations, and the poor nations are badly situated to hold them accountable. For another, the limited evidence on regional impacts suggests that it is the poorer nations that are most vulnerable to the worst impacts of climate change.[17] Finally, action on climate change creates a moral risk for the developed nations. It embodies a recognition that there are international norms of ethics and responsibility, and reinforces the idea that international cooperation on issues involving such norms is both possible and necessary. Hence, it may encourage attention to other moral defects of the current global system, such as global poverty, human rights violations, and so on.[18]

The Intergenerational Storm

We can now return to the three characteristics of the climate change problem identified earlier:

- Dispersion of causes and effects
- Fragmentation of agency
- Institutional inadequacy

The global storm emerges from a spatial reading of these characteristics, but I would argue that another, even more serious problem arises when we see them from a temporal perspective. I call this the "intergenerational storm."

Consider first the dispersion of causes and effects. Human-induced climate change is a severely lagged phenomenon. This is partly because some of the basic mechanisms set in motion by the greenhouse effect—such as sea-level rise—take a very long time to be fully realized. But it is also because by far the most important greenhouse gas emitted by human beings is carbon dioxide, and once emitted, molecules of carbon dioxide can spend a surprisingly long time in the atmosphere.[19]

Let us dwell for a moment on this second factor. The IPCC says that the average time spent by a molecule of carbon dioxide in the upper atmosphere is in the region of 5 to 200 years. This estimate is long enough to create a serious lagging effect; nevertheless, it obscures the fact that a significant percentage of carbon dioxide molecules remain in the atmosphere for much longer periods of time, on the order of thousands and tens of thousands of years. For instance, in a recent paper, David Archer says:

> The carbon cycle of the biosphere will take a long time to completely neutralize and sequester anthropogenic CO_2. We show a wide range of model forecasts of this effect. For the best-guess cases . . . we expect that 17–33% of the fossil fuel carbon will still reside in the atmosphere 1 kyr from now, decreasing to 10–15% at 10 kyr, and 7% at 100 kyr. The mean lifetime of fossil fuel CO_2 is about 30–35 kyr.[20]

This is a fact, he says, which has not yet "reached general public awareness."[21] Hence, he suggests that "a better shorthand for public discussion [than the IPCC estimate] might be that CO_2 sticks around for hundreds of years, plus 25% that sticks around for ever."[22]

The fact that carbon dioxide is a long-lived greenhouse gas has at least three important implications. First, climate change is a *resilient* phenomenon. Given that currently it does not seem practical to remove large amounts of emitted carbon dioxide from the atmosphere, or to moderate its climatic effects, the upward trend in atmospheric concentration is not easily reversible. Hence, a goal of stabilizing and then reducing carbon dioxide concentrations requires advance planning. Second, climate change impacts are *seriously backloaded*. The climate change that the earth is currently experiencing is primarily the result of emissions from some time in the past, rather than current emissions. As an illustration, it is widely accepted that by 2000 we had already committed ourselves to a rise of at least 0.5 and perhaps more than 1° Celsius over the then-observed rise of 0.6°C.[23] Third, backloading implies that the full, cumulative effects of our current emissions will not be realized for some time in the future. So, climate change is a *substantially deferred* phenomenon.

Temporal dispersion creates a number of problems. First, as is widely noted, the resilience of climate change implies that delays in action have serious repercussions for our ability to manage the problem. Second, backloading implies that climate change poses serious epistemic difficulties, especially for normal political actors. For one thing, backloading makes it hard to grasp the connection between causes and effects, and this may undermine the motivation to act;[24] for another, it implies that by the time we realize that things are bad, we will already be committed to much more change, so it undermines the ability to respond. Third, the deferral effect calls into question the ability of standard institutions to deal with the problem. For one thing, democratic political institutions have relatively short time horizons—the next election cycle, a politician's political career—and it is doubtful whether such institutions have the wherewithal to deal with substantially deferred impacts. Even more seriously, substantial deferral is likely to undermine the will to act. This is because there is an incentive problem: the bad effects of current emissions are likely to fall, or fall disproportionately, on future generations, whereas the benefits of emissions accrue largely to the present.[25]

These last two points already raise the specter of institutional inadequacy. But to appreciate this problem fully, we must first say something about the temporal fragmentation of agency. There is some reason to think that the temporal fragmentation of agency might be worse than the spatial fragmentation even considered in isolation. For there is a sense in which temporal fragmentation is more intractable than spatial fragmentation: in principle, spatially fragmented agents may actually become unified and so able really to act as a single agent, but temporally fragmented agents cannot actually become unified, and so may at best only act *as if* they were a single agent.

Interesting as such questions are, they need not detain us here. For temporal fragmentation in the context of the kind of temporal dispersion that characterizes climate change is clearly much worse than the associated spatial fragmentation. For the presence of backloading and deferral together brings on a new collective-action problem that adds to the tragedy of the commons caused by the global storm, and thereby makes matters much worse.

The problem emerges when one relaxes the assumption that countries can be relied on adequately to represent the interests of both their present and future citizens. Suppose that this is not true. Suppose instead that countries are biased toward the interests of the current generation. Then, since the benefits of carbon dioxide emission are felt primarily by the present generation, in the form of cheap energy, whereas the costs—in the form of the risk of severe and perhaps catastrophic climate change—are substantially deferred to future generations, climate change might provide an instance of a severe intergenerational collective-action problem. Moreover, this problem will be iterated. Each new generation will face the same incentive structure as soon as it gains the power to decide whether or not to act.[26]

The nature of the intergenerational problem is easiest to see if we compare it to the traditional prisoner's dilemma. Suppose we consider a pure version of the intergenerational problem, where the generations do not overlap.[27] (Call this the "pure intergenerational problem" (PIP).) In that case, the problem can be (roughly) characterized as follows:[28]

PIP1. It is *collectively rational* for most generations to cooperate: (almost) every generation prefers the outcome produced by everyone restricting pollution over the outcome produced by everyone overpolluting.

PIP2. It is *individually rational* for all generations not to cooperate: when each generation has the power to decide whether it will overpollute, each generation (rationally) prefers to overpollute, whatever the others do.

Now, the PIP is worse than the prisoner's dilemma in two main respects. The first respect is that its two constituent claims are worse. On the one hand, PIP1 is worse than PD1 because the first generation is not included. This means not only that one generation is not motivated to accept the collectively rational outcome, but also that the problem becomes iterated. Since subsequent generations have no reason to comply if their predecessors do not, noncompliance by the first generation has a domino effect that undermines the collective project. On the other hand, PIP2 is worse than PD2 because the reason for it is deeper. Both of these claims hold because the parties lack access to mechanisms (such as enforceable sanctions) that would make defection irrational. But whereas in normal prisoner's dilemma–type cases, this obstacle is largely practical, and can be resolved by creating appropriate institutions, in the PIP it arises because the parties do not coexist, and so seem unable to influence each other's behavior through the creation of appropriate coercive institutions.

This problem of interaction produces the second respect in which the PIP is worse than the prisoner's dilemma. This is that the PIP is more difficult to resolve, because the standard solutions to the prisoner's dilemma are unavailable: one cannot appeal to a wider context of mutually beneficial interaction, nor to the usual notions of reciprocity.

The upshot of all this is that in the case of climate change, the intergenerational analysis will be less optimistic about solutions than the tragedy of the commons analysis. For it implies that current populations may not be motivated to establish a fully adequate global regime, since, given the temporal dispersion of effects—and especially backloading and deferral—such a regime is probably not in *their* interests. This is a large moral problem, particularly since in my view the intergenerational problem dominates the tragedy of the commons aspect in climate change.

The PIP is bad enough considered in isolation. But in the context of climate change it is also subject to morally relevant multiplier effects. First, climate change is not a static phenomenon. In failing to act appropriately, the current generation does not simply pass an existing problem along to future people; rather it adds to it, making the problem worse. For one thing, it increases the costs of coping with climate change: failing to act now increases the magnitude of future climate change and so its effects. For another, it increases mitigation costs: failing to act now makes it more difficult to change because it allows additional investment in fossil fuel–based infrastructure in developed and especially less developed countries. Hence, inaction raises transition costs, making future change harder than change now. Moreover, and perhaps most importantly, the current generation does not add to the problem in a linear way. Rather, it rapidly accelerates the problem, since global emissions are increasing at a substantial rate—for example, total carbon dioxide emissions have increased more than fourfold in the last fifty years. Moreover, the current growth rate is around 2 percent per year.[29] Though 2 percent may not seem like much, the effects of compounding make it significant, even in the near term: "Continued growth of CO_2 emissions at 2% per year would yield a 22% increase of emission rate in 10 years and a 35% increase in 15 years."[30]

Second, insufficient action may make some generations suffer unnecessarily. Suppose that, at this point in time, climate change seriously affects the prospects of generations A, B, and C. Suppose, then, that if generation A refuses to act, the effect will continue for longer, harming generations D and E. This may make generation A's inaction worse in a significant respect. In addition to failing to aid generations B and C (and probably also increasing the magnitude of harm inflicted on them), generation A now harms generations D and E, who otherwise would be spared. On some views, this might count as especially egregious, since it might be said that it violates a fundamental moral principle of "Do no harm."[31]

Third, generation A's inaction may create situations where *tragic choices* must be made. One way in which a generation may act badly is if it puts in place a set of future circumstances that make it morally required for its successors (and perhaps even itself) to make other generations suffer either unnecessarily, or at least more than would otherwise be the case. For example, suppose that generation A could and should act now in order to limit climate change such that generation D would be kept below some crucial climate threshold, but delay would mean that they would pass that threshold.[32] If passing the threshold imposes severe costs on generation D, then their situation may be so dire that they are forced to take action that will harm generation F—such as emitting even more greenhouse gases—that they would otherwise not need to consider. What I have in mind is this. Under some circumstances actions that harm innocent others may be morally permissible on grounds of self-defense, and such circumstances may arise in the climate change case.[33] Hence, the claim is that, if there is a self-defense exception to the prohibition on harming innocent others, one way generation A might behave badly is by creating a situation such that generation D is forced to call on the self-defense exception and so inflict extra suffering on generation F.[34] Moreover, like the basic PIP, this problem can become iterated: perhaps generation F must call on the self-defense exception too, and so inflict harm on generation H, and so on.

The Theoretical Storm

The final storm I want to mention is constituted by our current theoretical ineptitude. We are extremely ill-equipped to deal with many problems characteristic of the long-term future. Even our best theories face basic and often severe difficulties addressing issues such as scientific uncertainty, intergenerational equity, contingent persons, nonhuman animals, and nature. But climate change involves all of these matters and more.[35]

Now I do not want to discuss any of these difficulties in any detail here. Instead, I want to gesture at how, when they converge with each other and with the global and intergenerational storms, they encourage a new and distinct problem for ethical action on climate change: the problem of moral corruption.

Moral Corruption

Corruption of the kind I have in mind can be facilitated in a number of ways. Consider the following examples of possible strategies:

- Distraction
- Complacency
- Unreasonable doubt
- Selective attention
- Delusion
- Pandering
- False witness
- Hypocrisy

Now, the mere listing of these strategies is probably enough to make the main point here, and I suspect that close observers of the political debate about climate change will recognize many of these mechanisms as being in play. Still, I would like to pause for a moment to draw particular attention to selective attention.

The problem is this. Since climate change involves a complex convergence of problems, it is easy to engage in manipulative or self-deceptive behavior by applying one's attention selectively, to only some of the considerations that make the situation difficult. At the level of practical politics, such strategies are all too familiar. For example, many political actors emphasize considerations that appear to make inaction excusable, or even desirable (such as uncertainty or simple economic calculations with high discount rates) and action more difficult and contentious (such the basic lifestyles issue) at the expense of those that seem to impose a clearer and more immediate burden (such as scientific consensus and the pure intergenerational problem).

But selective attention strategies may also manifest themselves more generally. And this prompts a very unpleasant thought: perhaps there is a problem of corruption in the theoretical, as well as the practical, debate. For example, it is possible that the prominence of the global storm model is not independent of the existence of the intergenerational storm, but rather is encouraged by it. After all, the current generation may find it highly advantageous to focus on the global storm. For one thing, such a focus tends to draw attention toward various issues of global politics and scientific uncertainty that seem to problematize action, and away from issues of intergenerational ethics, which tend to demand it. Thus, an emphasis on the global storm at the expense of the other problems may *facilitate* a strategy of procrastination and delay. For another, since it assumes that the relevant actors are nation-states who represent the interests of their citizens in perpetuity, the global storm analysis has the effect of assuming away the intergenerational aspect of the climate change problem.[36] Thus, an undue emphasis on it may obscure much of what is at stake in making climate policy, and in a way that may benefit present people.[37]

In conclusion, the presence of the problem of moral corruption reveals another sense in which climate change may be a perfect moral storm. This is that its complexity may turn out to be *perfectly convenient* for us, the current generation, and indeed for each successor generation as it comes to occupy our position. For one thing, it provides each generation with the cover under which it can seem to be taking the issue seriously—by negotiating weak and largely substanceless global accords, for example, and then heralding them as great achievements[38]—when really it is simply exploiting its temporal position. For another, all of this can occur without the exploitative generation actually having to acknowledge that this is what it is doing. By avoiding overtly selfish behavior, an earlier generation can take advantage of the future without the unpleasantness of admitting it—either to others, or, perhaps more importantly, to itself.

Notes

This chapter was originally written for presentation to an interdisciplinary workshop on Values in Nature at Princeton University, the proceedings of which appeared in *Environmental Values*. I thank the Center for Human Values at Princeton and the University of Washington for research support in the form of a Laurance S. Rockefeller fellowship. I also thank audiences at Iowa State University, Lewis and Clark College, the University of Washington, the Western Political Science Association, and the Pacific Division of the American Philosophical Association. For comments, I am particularly grateful to Chrisoula Andreou, Kristen Hessler, Jay Odenbaugh, John Meyer, Darrel Moellendorf, Peter Singer, Harlan Wilson, Clark Wolf, and two anonymous reviewers. I am especially indebted to Dale Jamieson.

1. Robert J. Samuelson, "Lots of Gain and No Pain!" *Newsweek* (February 21, 2005), 41. Samuelson was talking about another intergenerational issue—social security—but his claims ring true here as well.

2. Intergovernmental Panel on Climate Change (IPCC), *Climate Change 2001: Synthesis Report* (Cambridge: Cambridge University Press, 2001), 2; emphasis added.

3. For more on such issues, see Stephen Gardiner, "Ethics and Global Climate Change," *Ethics* 114 (2004): 555–600.

4. One might wonder why, despite the widespread agreement that climate change involves important ethical questions, there is relatively little overt discussion of these questions. The answer is no doubt complex. But my thesis may constitute part of that answer.

5. Sebastian Junger, *The Perfect Storm: A True Story of Men against the Sea* (New York: Norton, 1997).

6. This definition is my own. The phrase "perfect storm" is in wide use. However, it is difficult to find definitions of it. An online dictionary of slang offers the following: "When three events, usually beyond one's control, converge and create a large inconvenience for an individual. Each event represents one of the storms that collided on the *Andrea Gail* in the book/movie titled *The Perfect Storm*" (Urbandictionary.com, March 25, 2005).

7. An anonymous reviewer objects that this is a "very American take on the matter," since "the rest of the world" (a) "is less sure that there is an utter absence of effective global governance," (b) "might argue that were it not for recent U.S. resistance a centralized system of governance might be said to at least be in the early stages of evolution," and (c) accepts "Kyoto as a reasonable first step toward global governance on climate change." Much might be said about this, but here I can make only three quick points. First, suppose that (a)–(c) are all true. Even so, their truth does not seem sufficient to undermine the global storm; the claims are just too weak. Second, if there is a system of effective governance, then the current weakness of the international response to climate change becomes more, rather than less, surprising, and this bolsters one of my main claims in this chapter, which is that other factors need to be taken into account. Third, elsewhere I have criticized Kyoto as too weak (Gardiner, "The Global Warming Tragedy and the Dangerous Illusion of the Kyoto Protocol," *Ethics and International Affairs* 18 (2004): 23–39.). Others have criticized me for being too pessimistic here, invoking the "first-step" defense (e.g., Elizabeth Desombre, "Global Warming: More Common than Tragic," *Ethics and International Affairs* 18 (2004): 41-46.). My response is to say that it is the critics who are the pessimists: they believe that Kyoto was the best that humanity could achieve at the time; I am more optimistic about our capabilities.

8. The appropriateness of this model even to the spatial dimension requires some further specific, but usually undefended, background assumptions about the precise nature of the dispersion of effects and fragmentation of agency. But I will pass over that issue here.

9. Garrett Hardin, "The Tragedy of the Commons," *Science* 162 (1968): 1243–1248. I discuss this in more detail in previous work; see especially Stephen M. Gardiner, "The Real Tragedy of the Commons," *Philosophy and Public Affairs* 30 (2001): 387–416.

10. Nothing depends on the case being of this form. For a fuller characterization, see Gardiner, "The Real Tragedy of the Commons."

11. Some will complain that such game-theoretic analyses are misguided in general, and in any case irrelevant to the *ethics* of international affairs, since they focus on self-interested motivation. Though a full discussion is not possible here, a couple of quick responses may be helpful. First, I believe that often the best way to make progress in solving a given ethical problem is to get clear on what the problem actually is. Game theoretic analyses can sometimes be helpful here. (Some evidence for this is provided by their popularity in the actual literature on environmental issues in general, and climate change in particular). Second, my analysis need not assume that actual human individuals, states, or generations are exclusively self-interested, nor that their interests are exclusively economic. (In fact, I would reject such claims.) Instead, the analysis can proceed on a much more limited set of assumptions. Suppose, for example, that the following were the case: first, the actual, unreflective *consumption behavior* of most agents is dominated by their *perceived self-interest*; second, this is often seen in rather narrow terms; and third, such behavior drives much of the energy use in the industrialized countries, and thus much of the problem of climate change. If such claims are reasonable, then modeling the dynamics of the global warming problem in terms of a simplifying assumption of self-interest is not seriously misleading. For the role of that assumption is simply to suggest that (a) *if nothing is done to prevent it*, unreflective consumption behavior will dominate individual, state, and generational behavior, (b) this is likely to lead to tragedy, and so (c) some kind of regulation of normal consumption patterns (whether individual, governmental, market-based, or of some other form) is necessary to avoid a moral disaster.

12. This implies that, in the real world, commons problems do not strictly speaking satisfy all the conditions of the prisoner's dilemma paradigm. For relevant discussion, see Lee Shepski, "Prisoner's Dilemma: The Hard Problem," paper presented at the meeting of the Pacific Division of the American Philosophical Association, March 2006; Elinor Ostrom, *Governing the Commons: The Evolution of Institutions for Collective Action* (Cambridge: Cambridge University Press, 1990).

13. There is one fortunate convergence. Several writers have emphasized that the major ethical arguments all point in the same direction: that the developed countries should bear most of the costs of the transition—including those accruing to developing countries—at least in the early stages of mitigation and adaptation. See, for example, Peter Singer, *One World: The Ethics of Globalization* (New Haven, CT: Yale University Press, 2002); Henry Shue, "Global Environment and International Inequality," *International Affairs* 75 (1999): 531–545.

14. Rado Dimitrov argues that we must distinguish between different kinds of uncertainty when we investigate the effects of scientific uncertainty on international regime building, and that it is uncertainties about national impacts that undermine regime formation. See Rado Dimitrov, "Knowledge, Power and Interests in Environmental Regime Formation," *International Studies Quarterly* 47 (2003): 123–150.

15. This consideration appears to play a role in U.S. deliberations about climate change, where it is often asserted that the United States faces lower marginal costs from climate change than other countries. See, for example, Robert O. Mendelsohn, *Global Warming and the American Economy* (London: Edward Elgar, 2001); W. A. Nitze, "A Failure of Presidential Leadership," in Irving Mintzer and J. Amber Leonard, eds., *Negotiating Climate Change: The Inside Story of the Rio Convention* (Cambridge: Cambridge University Press, 1994); and, by contrast, National Assessment Synthesis Team, *Climate Change Impacts on the United States: The Potential Consequences of Climate Variability and Change* (Cambridge: Cambridge University Press, 2000), www.usgcrp.gov/usgcrp/nacc/default.htm.

16. Much more might be said here. I discuss some of the psychological aspects of political inertia and the role they play independently of scientific uncertainty in Stephen M. Gardiner, "Saved by Disaster? Abrupt Climate Change, Political Inertia, and the Possibility of an Intergenerational Arms Race," *Journal of Social Philosophy*, forthcoming.

17. This is so both because a greater proportion of their economies are in climate-sensitive sectors, and because—being poor—they are less able to deal with those impacts. See IPCC, "Summary for Policymakers," *Climate Change 2001: Impacts, Adaptation, and Vulnerability* (Cambridge: Cambridge University Press, 2001), 8, 16.

18. Of course, it does not help that the climate change problem arises in an unfortunate geopolitical setting. Current international relations occur against a backdrop of distraction, mistrust, and severe inequalities of power. The dominant global actor and lone superpower, the United States, refuses to address climate change, and is in any case distracted by the threat of global terrorism. Moreover, the international community, including many of America's historical allies, distrust its motives, its actions, and especially its uses of moral rhetoric, so there is global discord. This unfortunate state of affairs is especially problematic in relation to the developing nations, whose cooperation must be secured if the climate change problem is to be addressed. One issue is the credibility of the developed nations' commitment to solving the climate change problem. (See the next section.) Another is the North's focus on mitigation to the exclusion of adaptation issues. A third concern is the South's fear of an "abate and switch" strategy on the part of the North. (Note that considered in isolation, these factors do not seem sufficient to explain political inertia. After all, the climate change problem originally became prominent during the 1990s, a decade with a much more promising geopolitical environment.)

19. For more on both claims, see IPCC, *Climate Change 2001: Synthesis Report* (Cambridge: Cambridge University Press, 2001), 16–17.

20. David Archer, "Fate of Fossil Fuel CO2 in Geologic Time," *Journal of Geophysical Research*, 110 (2005), 5. "Kyr" means "thousand years."

21. David Archer, "How Long Will Global Warming Last?" March 15, 2005. http://www.realclimate.org/index.php/archives/2005/03/how-long-will-global-warming-last/#more-134.

22. Archer, "How Long Will Global Warming Last"; a similar remark occurs in Archer, "Fate of Fossil Fuels in Geologic Time." 5.

23. T. M. L. Wigley, "The Climate Change Commitment," *Science* 307 (2005): 1766–1769; Gerald Meehl, Warren M. Washington, William D. Collins, Julie M. Arblaster, Aixue Hu, Lawrence E. Buja, Warren G. Strand, and Haiyan Teng, "How Much More Global Warming and Sea Level Rise?", *Science* 307 (2005): 1769–1772; Richard T. Wetherald, Ronald J. Stouffer, and Keith W. Dixon, "Committed Warming and Its Implications for Climate Change," *Geophysical Research Letters* 28, no. 8 (2001): 1535–1538.

24. This is exacerbated by the fact that the climate is an inherently chaotic system in any case, and that there is no control against which its performance might be compared.

25. The possibility of nonlinear effects, such as in abrupt climate change, complicates this point, but I do not think it undermines it. See Gardiner, "Saved By Disaster?"

26. Elsewhere, I have argued that this background fact most readily explains the weakness of the Kyoto deal. See Gardiner, "The Global Warming Tragedy and the Dangerous Illusion of the Kyoto Protocol."

27. Generational overlap complicates the picture in some ways, but I do not think that it resolves the basic problem. See Stephen M. Gardiner, "The Pure Intergenerational Problem," *Monist* 86 (2003): 481–500.

28. These matters are discussed in more detail in Gardiner, "The Pure Intergenerational Problem," from which the following description is drawn.

29. James Hansen and Makiko Sato, "Greenhouse Gas Growth Rates," *Proceedings of the National Academy of Sciences* 101, no. 46 (2004): 16109–16114; James Hansen, "Can We Still Avoid Dangerous Human-Made Climate Change?", talk presented at the New School University, February 2006.

30. Hansen, "Can We Still Avoid Dangerous Human-Made Climate Change?", 9.

31. I owe this suggestion to Henry Shue.

32. See Brian C. O'Neill and Michael Oppenheimer, "Dangerous Climate Impacts and the Kyoto Protocol," *Science* 296 (2002): 1971–1972.

33. See Martino Traxler, "Fair Chore Division for Climate Change," *Social Theory and Practice* 28 (2002): 101–134 (see 107).

34. Henry Shue considers a related case in a recent paper (see Henry Shue, "Responsibility of Future Generations and the Technological Transition," in Walter Sinnott-Armstrong and Richard Howarth, eds., *Perspectives on Climate Change: Science, Economics, Politics, Ethics* (New York: Elsevier, 2005), 275–276).

35. For some discussion of the problems faced by cost-benefit analysis in particular, see John Broome, *Counting the Cost of Global Warming* (Isle of Harris, UK: White Horse Press, 1992); Clive L. Splash, *Greenhouse Economics: Value and Ethics* (London: Routledge, 2002); Gardiner, "Ethics and Global

Climate Change;"Gardiner, "Why Do Future Generations Need Protection?" working paper (Paris: Chaire Developpement Durable, 2006), http://ceco .polytechnique.fr/CDD/PDF/DDX-06-16.pdf.

36. In particular, it conceives of the problem as one that self-interested motivation alone should be able to solve, and where failure will result in self-inflicted harm. But the intergenerational analysis makes clear that these claims are not true: current actions will largely harm (innocent) future people, which suggests that motivations that are not generation-relative must be called on to protect them.

37. In particular, once one identifies the intergenerational storm, it becomes clear that any given generation confronts two versions of the tragedy of the commons. The first version assumes that nations represent the interests of their citizens in perpetuity, and so is genuinely cross-generational, but the second assumes that nations predominantly represent the interests of their current citizens, and so is merely intragenerational. The problem is then that the collectively rational solutions to these two commons problems may be—and very likely are—different. (For example, in the case of climate change, it is probable that the intragenerational problem calls for much less mitigation of greenhouse gas emissions than the cross-generational problem.) Thus, we cannot take the fact that a particular generation is engaged in resolving the first version (the intragenerational tragedy) as evidence that they are interested in solving the second (the cross-generational version). See Gardiner, "The Global Warming Tragedy and the Dangerous Illusion of the Kyoto Protocol."

38. Gardiner, "The Global Warming Tragedy and the Dangerous Illusion of the Kyoto Protocol."

3

Climate Change, Environmental Rights, and Emission Shares

Steve Vanderheiden

The 1992 UN Framework Convention on Climate Change (UNFCCC) declared anthropogenic climate change to be a "common concern of mankind" and resolved to take all necessary steps in order to prevent "dangerous anthropogenic interference with the climate system." Noting that "the largest share of historical and current global emissions of greenhouse gases has originated in developed countries," the 192 national signatories to the treaty pledged to freeze greenhouse gas (GHG) emissions at 1990 levels by the year 2000 and, through future international action undertaken through the auspices of the UNFCCC process, to "protect the climate system for the benefit of present and future generations of mankind, on the basis of equity and in accordance with their common but differentiated responsibilities and respective capacities." It was for these latter two explicitly recognized reasons—that the world's industrialized countries were primarily responsible for causing the problem and were uniquely capable of its mitigation—that concerns for "equity" were held to require that "the developed countries take the lead in combating climate change and the adverse effects thereof."[1] Five years later, the Kyoto Protocol attempted to translate those commitments into policy, imposing binding emission caps on the world's industrialized nations while temporarily postponing the imposition of such caps on developing countries, and in doing so provided the rationale for both the U.S. Senate's initial opposition to the treaty's ratification (declared 95–0 with the 1997 Byrd-Hagel Resolution) and the George W. Bush administration's later formal withdrawal from the Kyoto framework in 2001. Calling the treaty "unfair," the administration echoed the Senate's earlier claim that the "disparity of treatment" between industrialized nations and developing countries like China and India (both specifically mentioned) justified U.S. nonparticipation in that climate policy regime.

Indeed, normative concerns for fairness have featured prominently throughout the global climate policy process, and debates over the treaty's fairness are inseparable from those about its efficacy, because no unfair global climate regime stands a chance of gaining the requisite assent of the world's nations and no ineffective agreement can mitigate the unfairness of an environmental problem that is disproportionately caused by the world's affluent while expected to visit disproportionate harm on the world's poor.[2] But can a global climate regime be simultaneously fair and effective? What would such a regime look like? No climate policy regime can be effective unless it limits global GHG emissions to some level substantially below present rates, and the primary task of a global climate change mitigation effort involves the allocation of emission caps along two dimensions: among nations[3] and over time. While the latter distributive problem involves the determination of some maximum allowable aggregate level of annual emissions—where higher current emissions necessarily entail lower future ones given any future atmospheric GHG concentration target—the former allocates this annual total among nations. With these two distributive problems at the core of global climate policy, how are fairness and equity to be conceived, how are they related, and how can these aims simultaneously be achieved? In this chapter, I argue that these dual imperatives can best be conceived through the critical examination of three key environmental rights, which together inform the most defensible formula for allocating national emission shares. Before considering the claims made under these three varieties of rights, though, we might first consider their relation to issues of fairness from a developing-country perspective on the allocation of national emission shares, for both elucidation and illumination of the various rights-based claims contained there.

In response to the U.S. government's claim that the Kyoto Protocol is unfair, the Delhi-based Centre for Science and Environment (CSE) invokes a counterargument about fairness. It rejects the premises implicit in the Bush/Senate claim and highlights two kinds of wide global disparities between affluent industrialized nations like the United States and poor developing ones like India, arguing that such disparities constitute morally relevant differences that justify their differential treatment under a global climate regime:

The total carbon dioxide emissions from one U.S. citizen in 1996 were 19 times the emissions of one Indian. U.S. emissions in total are still more than double those from China. At a time when a large part of India's population does not even have access to electricity, Bush would like this country to stem its "survival emissions," so that industrialized countries like the U.S. can continue to have high "luxury emissions." *This amounts to demanding a freeze on global inequity, where rich countries stay rich, and poor countries stay poor, since carbon dioxide emissions are closely linked to GDP growth.*[4]

Several normative claims about fairness are presented here, and each is worth considering with some care, because each bears on the fair allocation of the requisite costs of a global climate regime. First, the CSE argument distinguishes between a basic minimum level of individual GHG emissions that all persons need in order to survive (termed "survival emissions") and those that go beyond what are necessary for mere survival, resulting instead from activities usually associated with affluence ("luxury emissions"). The CSE position is that the former ought to be granted a higher priority than the latter such that survival emissions can never be limited in order to allow more luxury emissions. Second, the argument claims that *excessive* emissions (not all emissions) are what should be seen as causally responsible for the problem—implying that anthropogenic climate change would not exist if all persons emitted GHGs at the average Indian per capita rate—so that luxury but not survival emissions connote moral responsibility (in the form of liability) for redressing it through a global climate regime. Finally, it asserts a right for nations like India to develop (based in concerns for global equity to be surveyed below), increasing their per capita GHG emissions well beyond the threshold defining survival emissions at the same time that industrialized nations that were assigned emission caps under the Kyoto Protocol are required to decrease their overall and per capita emissions.

In addition to the explicitly claimed right to develop, another kind of right, and one that is likewise grounded in an equity-based conception of fairness, is implicitly asserted above. Insofar as a global climate regime must be charged with allocating the remedial costs of GHG emission reductions—whether through national emission caps, compensatory liability, or both—assessments of liability ought to be fault-based rather than relying on a standard of strict liability, such that nations are held liable only for further emissions beyond the "survival" threshold. By this line of argument, no person can be faulted for acts necessary for survival, because they cannot plausibly be expected to refrain from committing them—or as Kant famously put the same point, *ought* implies *can*—and such persons (or the national governments that represent them in a global climate regime) would likely lack the remedial capacity necessary for assuming liability without fault, since their emissions cannot by definition be reduced any further without impairing basic functioning. In assigning national emission caps or in otherwise assessing remedial liability for climate change, then, the argument asserts a right to some basic minimum level of per capita GHG emissions, below which nations cannot be held responsible for causing climate change (and so do not deserve to be assigned liability for its mitigation), but above which they begin to incur liability for their respective causal contributions to climate change. Survival emissions, in other words, are protected by right and are viewed as faultless, while liability is assessed based on each nation's luxury emissions, which are faulty and to which none have a right. India, by this argument, cannot be assigned liability for causing climate change—at least insofar as its average citizen produces only survival emissions—but the United States (where per capita emission rates are nineteen times higher and are substantially above survival rates) can be faulted for causing the problem, and must consequently pay the costs of its mitigation.

These two kinds of rights claims (along with a third to be introduced below) together inform the design of a fair and effective global climate regime, because they provide a normative framework for distributing emission shares along the two dimensions noted above. First and most obviously, the claim to some share of the atmosphere's *absorptive capacity* (the common-pool resource that allows a finite quantity of GHGs to be safely absorbed into terrestrial sinks without raising atmospheric concentrations of those heat-trapping gases) can be regarded as a kind of right—captured in the asserted right to survival emissions—and one implicitly invoked each time people engage in the myriad activities that produce such gases. Since this ecological capacity is finite, the assignment of emission rights must likewise be capped at some level; the distinction between survival and luxury emissions implies that per capita emission cap be set at a level equal to or above the survival threshold, and (assuming a priority for the former) that none be allowed luxury emissions until all are guaranteed survival emissions. Should this per capita emission cap be set too high, however (where, as is currently the case, total annual GHG emissions exceed absorptive capacity), a second right (to *climatic stability*, held by both current and future generations) comes into play. Luxury emissions must be strictly limited, or increasing atmospheric GHG concentrations will cause significant climatic instability, likely producing the range of adverse effects identified by climate scientists and thereby violating the rights of current and future persons to a stable climate. On the other hand, assigning emission caps to developing countries that are too stringent to allow for their development (as would be the case if countries like India or China were held to the historical baseline formula of the Kyoto Protocol) raises the possibility of violating a third right (one to *development*). While this latter right may at first appear to be distinct in kind from the first two (because it ostensibly reflects economic interests rather than those of environmental protection), the case for development rights can be cogently understood as claims to environmental space: nations or persons must be allowed adequate atmospheric absorptive capacity not only for mere survival (as is supposed by the first rights-based claim), but may also have a claim to sufficient GHG allowances or luxury emissions to allow for economic or human development, and therefore for human flourishing.

I weigh these rights-based claims below, paying attention to the way each might be normatively justified as well as to the implications of each for the design of a fair and effective global climate regime, concluding not only that all three of these rights-based claims are valid (i.e., they represent important interests and so fit within existing schemes of similar rights in both structure and justification), but that they together imply a rather specific allocation formula for the manner in which global emission shares are assigned to nations or persons. That is, a just global emission allocation is one that (1) pays sufficient attention to global emission caps such that it avoids causing future climatic instability; (2) ensures that the distribution of emission shares among and within nations allows for adequate economic and human development; and (3) assigns the remedial costs associated with climate change mitigation in accordance with a defensible account of moral responsibility, in which fault-based national liability is assigned in accordance with luxury but not survival emissions (because one cannot be faulted for claiming some share of a common resource to which one is entitled as a matter of right). If all three of these rights are recognized and protected, a global climate regime will be required to allocate emission shares much more equitably than is the case under status quo use-based claims, where the richest 20 percent of the world now makes de facto claims on the vast majority of atmospheric space, or even under those schemes (like the Kyoto Protocol) that have been developed under the auspices of the UNFCCC. Such considerations are not merely of theoretical interest: allocating GHG emission shares in a fair or rights-protecting manner is an essential feature of any effective global climate regime, since any effective global regulatory apparatus must necessarily rely on the voluntary cooperation of member nations—subject to binding caps and with sufficient mechanisms for monitoring and ensuring compliance—and no nation can voluntarily submit to terms that violate the rights of its citizens (present or future). Recognizing such environmental rights and building that recognition into the structure of a global climate regime is therefore both a principled and practical project, and one with some current urgency.

Environmental Rights

While commitments to ideals of equity and responsibility are declared in the text and entrenched in the design of the climate convention,[5] their formal recognition in law and policy can more effectively be accomplished by substantiating them as environmental rights, which provide the necessary legal and political support for assisting right holders in having their claims recognized. Since a right connotes a valid claim to either provision (in the case of positive rights) or noninterference (with negative rights), the formal legal protection of rights offers a more robust form of protection for the interests they are designed to advance—providing potential claimants an avenue of appeal against rulings or shortcomings of the climate regime, allowing for a quasi-judicial check on its administration—as well as having a powerful effect on the formation of social norms. Tim Hayward notes that instantiating aims of environmental protection in legal or constitutional rights "entrenches a recognition of the importance of environmental protection; it offers the possibility of unifying principles for legislation and regulation; it secures these principles against the vicissitudes of routine politics, while at the same time enhancing possibilities of democratic participation in environmental decision-making processes."[6] Given the trumping power of constitutional rights, Hayward makes the case for them by comparing environmental rights to the set of recognized universal human rights, noting the intuitively plausible claim that "an adequate environment is as basic a condition of human flourishing as any of those that are already protected as human rights."[7] Hayward argues that the right to an adequate environment meets the standard test for a genuine human right, since it protects human interests that are "of paramount moral importance" (given that "environmental harms can threaten vital human interests"), and that such a right would also be genuinely universal, because "the interests it is intended to protect are common to all humans."[8]

Although several formulations of environmental rights exist in law and in the academic literature on human rights, the most general and encompassing formulation of the range of interests they might protect posits a right to a physical environment that provides the material basis for human flourishing (and not merely survival—a point to be considered further below), an exemplary version of which can be found in the opening principle of the Stockholm Declaration, negotiated at the 1972 UN Conference on the Human Environment:

Principle 1: Man has the fundamental right to freedom, equality, and adequate conditions of life, in an environment of a quality that permits a life of dignity and well-being, and he bears a solemn responsibility to protect and improve the environment for present and future generations.[9]

Essential to the expression of the right to an adequate environment are several features that are instructive to the case for a right to climatic stability, which I consider below. First, the environmental conditions that the right aims to secure are set alongside basic human ideals of freedom and equality in order to emphasize that all three (not merely the first two) ought to have the status of fundamental rights, which have priority over other less important rights. The three are interrelated—that is, in the absence of adequate environmental conditions, the ideals of freedom and equality cannot fully be realized, and greater freedom and equality may also be necessary conditions for protecting the environment. Second, all three of these rights are associated with human dignity and well-being, which aims to undercut the common assertion that environmental protection trades off against human welfare. Finally, the resolution associates this right to a "solemn responsibility" or correlative duty of environmental protection to which persons are obligated from cosmopolitan and intergenerational justice.

The right to an adequate environment is intended to encompass a broad range of duties of environmental protection in which persons are the primary beneficiaries of obligatory actions,[10] and the right to climatic stability appears to be an obvious corollary of such a right. While climate change is only one of many ongoing threats to the maintenance of an adequate environment, it must be regarded as among the most serious threats. Therefore, meeting one's duty to maintain climatic stability— requiring limits on excessive GHG emissions—is a necessary but insufficient condition for satisfying the general obligation to maintain an adequate environment, making the right to climatic stability a subsidiary right to the general right sketched above. Whether the right to climatic stability is sufficiently distinct or inherently weighty to require a separate legal or constitutional mention, or whether by contrast it ought to be considered as a necessary part of a more general fundamental right to an adequate environment, need not concern us here. Suffice it to say that the two are very closely associated and share a similar form, so that the case for the more general right for which Hayward argues entails the recognition (in some form) of a right to climatic stability. Before we can fully understand the implications of recognizing the right to an adequate environment, though, we must first consider the two rights against which its GHG-reducing imperatives must be balanced, because both the right to survival emissions (or, more generally, to subsistence) as well as the right to develop both weigh in the opposite direction. Both are also jeopardized by global emission allocation schemes that excessively limit national emissions in order to guard against climatic instability.

GHG Emission Rights

Rather than beginning with the more ambitious claim that residents of developing nations have a right to develop—or, in what amounts to the same thing, that all persons are entitled to some level of luxury emissions—I will begin with the more modest claim that all persons have a basic or fundamental right to survival emissions. In examining a nation's historical and current emissions, such a right requires that we distinguish between survival and luxury emissions, where none can be held liable for climate-related harm that results from the former, but where all must be assigned remedial costs in proportion to their share of the latter. By this distinction, survival emissions clearly warrant the status of a basic right—for reasons explored below—but luxury emissions do not so clearly qualify for the sort of protection that rights entail, since the interests they represent are less basic than those of survival emissions. Moreover, persons can be held responsible for their luxury emissions in a way that they cannot be held accountable for their survival emissions, following Brian Barry's formulation of the *principle of responsibility*, which holds that "a legitimate origin of different outcomes for different people is that they have made different voluntary choices. . . . The obverse of this principle is that bad outcomes for which somebody is not responsible provide a prima facie case for compensation."[11] In no sense can the emissions that a person produces as minimally necessary to meet their basic needs be attributed to voluntary acts or choices—they are, by definition, unavoidable—but persons can and do elect to produce emissions beyond that threshold. By this principle, therefore, persons must be held responsible for the latter and not for the former, as through assessments of liability for climate-related harm.

Insofar as all persons have a vital interest in the necessary conditions for human survival, they have a vital interest in survival emissions; insofar as they have a somewhat less basic, but still very important, interest in flourishing, they have a strong (if not vital) interest in further emissions beyond those minimally necessary for survival. We might put this observation in a slightly different way in order to illuminate the role of rights in the design of a global climate regime: persons have a *basic right* (i.e., a strong claim of entitlement capable of trumping nonbasic interests and with associated claims for legal remedy to deprivations of those interests related to the right) to their survival emissions, but they have only a nonbasic right (or a weaker claim that is trumped by stronger ones) to their luxury emissions. Several implications follow from formulating this distinction in terms of rights. As previously observed, persons have valid claims of entitlement to emit GHGs up to the survival threshold, so assessments of liability (which imply fault) cannot be made against acts to which persons are entitled as a matter of right. Additionally, this right entails a valid claim for remedial state assistance when others (whether states or private parties) threaten the practical ability of persons to exercise the right—for example, a climate regime may be required to curb luxury emissions in order to allow sufficient atmospheric space for survival emissions. Finally, survival emissions maintain their priority over luxury emissions even in the context of significant global inequality, since their status as a basic right makes the claim to survival emissions trump claims based on lesser rights, such as the property-right claims usually wielded in defense of inequality. Thus, the government of a poor nation cannot be allowed to sell "unused" survival emissions in GHG markets to a rich nation seeking additional luxury emissions, regardless of the price offered in return.

On what basis should a right to survival emissions be regarded as alongside the widely recognized set of universal human rights to which all are entitled? Perhaps the most compelling case for this kind of environmental right follows from Henry Shue's work on subsistence rights. Shue argues against the conventional distinction between security rights (which are often regarded as more fundamental and so are better protected under law) and economic rights (which enjoy considerably less protection), suggesting that the more salient distinction among categories of rights—and one supporting a priority system for weighing competing rights claims—is between basic and nonbasic rights. Basic rights, he suggests, "specify the line beneath which no one is to be allowed to sink" and so constitute "everyone's minimum demand upon the rest of humanity." A right that is genuinely basic trumps the rights that may assist in the further development of human potential but that are not essential for basic functioning in cases where the two kinds of rights conflict, Shue argues, and ought to be given greater protection when deploying scarce resources. The justification for this priority is built into the idea of a basic right itself, he contends: "When a right is genuinely basic, any attempt to enjoy any other right by sacrificing the basic right would be quite literally self-defeating, cutting the ground from beneath itself."[12]

Although security rights (e.g., rights against harm, wrongful arrest, or excessive punishment) have long been enshrined within law and protected by states, social and economic rights (e.g., the right to public education or to organize a labor union) have more recently begun to be added to legal and political documents alongside these security rights, engendering at least some controversy about whether they belong there. As the standard criticism goes, security rights are more fundamental than are social or economic rights in that they protect the most basic human interest not to be harmed—a person can survive without a public education or a labor union, but not without basic protection from harm—and for this reason ought to enjoy priority over less important rights. Societies may elect to provide social and economic rights after all security rights have been provided for, this position maintains, but the former are optional and clearly of a lesser priority. Moreover, the standard criticism equates security rights with negative rights, which are more cheaply and easily provided by states since negative rights correspond only with the duty to refrain from certain acts, whereas social and economic rights are usually equated with positive rights, which are assumed to be more costly to provide. To illustrate this distinction as it is often made, if you have a negative right (e.g., against harm), then I have a correlative duty (relatively easily met) to refrain from harming you, but if you have a positive right (e.g., to public education), then I have a more demanding correlative duty to help pay for it through my taxes. Taken together, the standard criticism maintains that new economic or social rights should not be added to the already lengthy lists of individual or human rights until security rights have been fully guaranteed, and then only after considering the opportunity costs of providing for such expensive claims.

Against these prevailing distinctions between economic and security rights, Shue points out that many security rights are partially positive in character, requiring provision rather than mere restraint, and are actually quite expensive to maintain, because they include the costs of domestic law enforcement and the military, along with the judicial and penal systems, while many economic rights (e.g., antipollution regulations and workplace safety standards) are largely negative and cost relatively little for the state to provide. Insofar as the conventional case against economic rights rests on the mistaken assumption that they are more costly for a state to guarantee, Shue offers the basic-nonbasic distinction as a more defensible priority system for weighing conflicting claims and in deploying scarce state resources. Rather than relying on false generalizations (e.g., that all security rights are negative and thus inexpensive to maintain) married to cost-benefit analysis (which supposes that protecting rights is only justified when it is cost-effective for the state to do so), Shue argues for a priority system based on the extent to which rights protect activities that are essential to meeting human needs or safeguard those that are instrumental to human flourishing. According to Shue, the protection of basic rights is a matter of basic justice, and ought therefore to be secured before attempts are made to provide for less basic rights.

Once we think of rights in this way, a basic right to survival emissions becomes plausible. As Shue notes, basic rights protect vital human interests in physical security as well as minimal economic security (or *subsistence*), with the latter including "unpolluted air, unpolluted water, adequate food, adequate clothing, adequate shelter, and minimal preventative health care."[13] In contrast to global efforts that have treated poverty and hunger as problems of charity rather than issues of justice or basic rights, Shue urges that subsistence be socially guaranteed as a matter of right, or else "attempts to actually enjoy the other rights remain open to a standard threat like the deprivation of security or subsistence."[14] Moreover, since the priority of basic over nonbasic rights is based on the parallel distinction between vital and nonvital interests, Shue argues that the world's affluent are required to sacrifice (in increasing order of importance) their preference satisfaction, cultural enrichment, and nonbasic rights—and are permitted, but cannot be required, to sacrifice some of their basic rights—in order to secure the basic rights of the world's poor.

According to Shue, this obligation of justice is based on the *vital interests principle*, which holds that "it is unfair to demand of people actions the very performance of which would preclude for themselves a way of protecting a vital interest while failing to provide some other protection for that interest, when it is possible to protect it by means that do not threaten the vital interests of anyone."[15] Failing to act in order to protect threatened basic rights when this can be accomplished without the sacrifice of anyone's basic rights is essentially to fail to regard persons as equals; it is to give priority to the nonvital interests of some over the vital interests of others. It is, in other words, a violation of a fundamental premise of egalitarian justice: that no person is intrinsically more valuable than another. For Shue, this entails a duty on the part of national governments, as uniquely capable of securing basic rights globally and as "powerful institutions capable of causing severe deprivations when they do not restrain themselves," to avoid depriving others of their basic rights, whether through a negative act of restraint—for example, by avoiding causing catastrophic environmental problems like climate change—or positive acts of provision.

In addition to various basic rights that Shue lists above, one might also suppose that a stable climate might be considered as among the basic rights of humans (as Hayward claims), and that anthropogenic climate change therefore threatens not only to transfer substantial costs onto the world's poor—indirectly affecting their subsistence rights—but also to undermine their subsistence by reducing crop yields, threatening water availability and quality, and in some cases threatening the territorial integrity of entire peoples. This latter case may be the most compelling of all rights issues related to climate change, because residents of low-lying and small island states have recently become increasingly vocal in climate debates by invoking their rights against having large parts (even all, in some cases) of their current territories inundated by projected sea-level rises or being displaced by vanishing tundra. In 2005, for example, representatives of the Inuit (a people of 155,000 residing in Arctic regions of Canada, Alaska, Greenland, and Russia) filed a petition with the Inter-American Commission on Human Rights alleging that the United States, in its capacity as primary obstacle to an effective global climate regime as well as biggest GHG polluter, was violating their human rights by exacerbating global warming, since the warming trends caused by increasing GHG concentrations have already produced profound thinning effects on Arctic ice sheets, threatening species on which Inuit hunters depend and so, in effect, threatening the cultural preservation of the entire Inuit people. Since climate change is widely expected to threaten wildlife and shift species habitats, myriad other potential threats to traditional cultures may likewise be affected in similar fashion—threatening the preservation of indigenous culture, if not territorial integrity—thus elevating the prominence of such rights claims in the climate debate.

In fact, the idea of a right to survival emissions finds some direct theoretical mention in Shue's work on subsistence rights, which may also lend support to the notion of a right to develop on the part of poor countries that now have difficulty meeting the basic needs of their citizens (I consider an extension of this idea below). Classifying access to the atmosphere's "emission absorptive capacity" as among basic rights, he suggests that persons ought to be entitled to a basic minimum level of per capita GHG emissions as a matter of right (essentially describing the concept of survival emissions), which entails that provision of this entitlement ought therefore to trump the exercise of other nonbasic rights: "For practically everyone at present, and for the immediate future, survival requires the use of GHG emission absorptive capacity. No reasonable, immediate alternative exists. Strange as it might initially sound, emission absorptive capacity is as vital as food and water and, virtually everywhere, shelter and clothing."[16] Insofar as basic or subsistence rights can be understood as protecting vital interests, where one cannot enjoy other rights unless these basic rights are first protected, then supposing that survival emissions count among a person's basic rights clearly follows. Since persons literally cannot survive without them, they are as basic as physical security and the standard set of subsistence rights (rights to food, shelter, clean water, and so on). Moreover, as Shue suggests, attributing basic rights to survival emissions generates a distributive principle that applies to the assignment of national emission shares: that "the only morally permissible allocations of emissions are allocations that guarantee the availability of the minimum necessary emissions to every person, which entails reserving adequate unused absorptive capacity for those unused emissions."[17]

Development Rights

If we suppose that nations like India have a right to develop, then we must also suppose that they have a right to emit GHGs at a per capita level considerably above the level of survival emissions, and one much closer to those now granted to industrialized nations. The argument made on behalf of such a right, and a recurrent one within the debates surrounding the development of the climate convention, invokes the normative ideal of equity, which is to be applied either to living standards, GHG emissions, or both. When so applied, it becomes clear what this claim is asserting: that the current worldwide distribution of wealth is highly inequitable (and is reflected in highly unequal national emission rates), that justice demands that these inequities be reduced, and that imposing emission caps on developing nations like India at levels too low to allow for their industrialization and other forms of economic development would in effect freeze the world's nations in their present state of development, allowing rich countries to continue producing per capita GHG emissions at rates far higher than those in developing nations, unfairly reserving the benefits of high emission allowances for nations currently among the world's affluent, and preventing their spread elsewhere. But on what basis might such a right be justified?

As Thomas Athanasiou and Paul Baer point out, assigning equitable emission caps to developing nations need not be justified on principled grounds alone, since "a climate treaty that indefinitely restricts a Chinese (or Indian) to lower emissions than an American (or European) will not be accepted as fair and, finally, will not be accepted at all."[18] Neither India nor China would be able to accept any climate convention that assigned them per capita emission caps too low to allow for industrialization or increasing consumption rates, since these would constitute a de facto barrier to development. On the other hand, no climate regime that excluded India and China from GHG emission caps altogether stands a chance of arresting the current growth in global emissions. Normative concerns based in the UNFCCC's "common but differentiated responsibilities and respective capabilities" model may have been the stated justification for the decision to exempt developing nations from mandatory caps during the Kyoto Protocol's first compliance period (ending in 2012). But also significant was the desire on the part of industrialized countries to defer discussion of the incendiary question of how high or low to set any assigned per capita GHG caps in India and China if they were to be included under the first round of binding caps. Resolving this problem to the satisfaction of both industrialized and developing nations would have been far more difficult than simply deferring the question entirely (as was accomplished by exempting the latter from any mandatory caps), since basing future caps for China or India on their 1990 baseline emissions would have raised valid objections about denying their rights to develop.

Developing countries could not have been assigned and would not have accepted caps that represented per capita emission rates that amount to a mere fraction of those allowed within industrialized nations—as would be the case if they were indexed to the 1990 baseline—but neither could they be assigned caps set at the levels allowed for Europe and

Japan (let alone the United States), for this would allow for significant increases in worldwide emissions even with developed nations meeting their assigned targets. In China, where there are 8 motor vehicles for every 1,000 people, or in India, where there are 7, mandatory emission caps comparable to those assigned to the United States (where there are 767 inefficient automobiles for every 1,000 people) would obviously be unfair. As Athanasiou and Baer note, "We cannot hope to find justice in a world where the poor come to live as the rich do today, for there is not world enough. There will have to be some other kind of solution. There will, indeed, have to be new dreams on all sides, and the rich, in particular, will have to make those dreams possible by learning to share."[19] Ecological limits on atmospheric absorptive capacity require that increases in allowable per capita GHG emissions from developing nations like India and China be accompanied by even larger decreases in allowable per capita emissions in industrialized nations (with their smaller populations) merely in order to freeze global emissions—the assignment of emission shares among nations from a fixed aggregate cap being a zero-sum game—to say nothing of the emission cuts that would be necessary if humanity is to fulfill the UNFCCC's mandate of avoiding dangerous interference with the planet's climate system.

Denying developing countries sufficient GHG emission allowances in order to accommodate development would have been tremendously unfair and unacceptable to those countries, but adjusting the assigned emission allowances among the industrialized nations to reflect significant per capita increases in India and China while allowing the same global aggregate emission levels would have been hugely unpopular, and even less likely to be accepted by the relevant parties. That is, opening up the question of developing-country emission caps would have forced delegates attending the conventions to at least consider the "contraction and convergence" scenario advocated by many climate activists, which would have required the industrialized nations to significantly reduce their GHG emissions (the *contraction*, with emission cuts much steeper than those prescribed under the Kyoto Protocol) in order to allow developing nations to eventually be allowed per capita emission caps equal to those in the industrialized nations (the *convergence*). As Athanasiou and Baer suggest, ecological limits will not allow for convergence (a demand of equity if developing countries are allowed the right to develop) without contraction (a necessary condition for the climate regime's efficacy). During the UNFCCC process, temporarily exempting developing nations from emission caps seemed a more plausible strategy than trying to convince either the developing countries to accept limits on development or industrialized ones to accept contraction—an estimation that has only been underscored by the U.S. refusal to accept even modest cuts without similar caps being applied to developing countries.

Exempting developing countries from the initial round of binding caps and timelines may have been politically expedient, but is it unfair to industrialized countries like the United States, as the Senate and the Bush administration have alleged? Here, it is helpful to think of a climate regime as essentially a decision about how to allocate costs for mitigation, adaptation, and compensation, and to rely on theories of justice and responsibility for assistance in determining a fair distribution of costs. As noted above, the principle of responsibility (as seen in fault-based liability) offers an account of the proper distribution of fault-related costs: those who are responsible for causing the problem through their historical emissions are the ones that should pay, and in proportion to their historical luxury (but not survival) emissions. Despite being home to 40 percent of the planet's population, China and India have together contributed only 9 percent of the total quantity of accumulated anthropogenic GHGs, compared with over 30 percent by the United States, which contains less than 5 percent of the world's population. By a standard of strict liability (making no distinction between survival and luxury emissions), the United States should bear 30 percent of total remedial costs, with China and India bearing a combined 9 percent. As the Framework Convention acknowledges, the world's industrialized countries—which are together responsible for over 75 percent of historical GHG emissions though comprising only 20 percent of the world's population—bear primary responsible for causing global climate change, and so (by a standard of strict liability) ought to be assigned at least that same share of liability for its remedy. But should they accept *all* of the liability for the problem, if that is the effect of developing-country exemptions? Even though the average Indian produces significantly less climate-changing gas than does the average American, they still contribute some GHGs into the atmosphere. Shouldn't they accept some share of the costs, perhaps in proportion to India's total historical emissions?

Maybe or maybe not, depending on which version of the "survival emissions" claim is most defensible. A weak version of it holds Indians to be significantly less responsible (but still responsible) for causing climate change, thus requiring them by the principle of responsibility to bear some of the costs of its remedy, while a strong version (and the one implied by the CSE argument) would maintain that Indians are not responsible at all for causing climate change, and therefore ought to be assigned no remedial burdens. Assuming that average Indians produce only survival and not luxury emissions (an assumption pursued below), the weak version is based on a straightforward application of strict liability, assigning remedial costs in proportion to historical GHG emissions, but how might a defense of the strong version be formulated? The strong version posits a basic minimum level of emissions to which persons might be entitled without being assessed liability, and assigns national liability based on the total amount of historical emissions emitted above this minimum—mitigation costs assigned to each nation would be based on its historical share of total luxury emissions. Insofar as developing countries like India and China have historically produced only survival emissions and not luxury emissions, they bear no responsibility for causing climate change, and hence can be attributed no liability for its remedy. The argument for the strong version thus depends on the association between fault and causal responsibility for avoidable harm, or (to put the same point another way) Barry's principle of responsibility, which attributes fault and assigns liability only to emissions associated with activity beyond that necessary for bare subsistence. Either way, still lacking is an argument for granting persons a right to some quantity of luxury emissions as necessary for development, given that both of these two standards assume liability for climate-related harm to be based on emissions that exceed survival emissions, and such assessments may interfere with the practical realization of a right to develop.

Recall that higher per capita emission allowances for developing countries entail the "contraction and convergence" scenario of more equitably allocated emission shares among the world's peoples and persons, because finite atmospheric space necessitates a zero-sum game where per capita emission increases for some are possible only with offsetting emission cuts for others. The argument for a right to survival emissions can be grounded (as Shue argues) in basic subsistence rights, but the right to develop requires a more difficult case from egalitarian justice, rather than one simply based in the avoidance of harm or protection of basic rights. Within the distributive justice literature, this contrast is often referred to as between an equal distribution and one guaranteeing a basic minimum, and can be found, for example, in the claim that justice only demands that all persons have access to a basic minimum set of social resources, not that the resources be equally allocated. The case for a universal human right to some level of survival emissions is more easily made than the case for a human right to development, but the latter is nonetheless asserted above and now stands in need of some justification. Having surveyed the case for the more limited right defended by Shue, and based in the distinction between survival and luxury emissions, how might the latter argument (which presumably allows for considerably more luxury emissions for those in India and considerably fewer for those in the United States) go?

From classical liberalism, a long-standing constraint on inequality comes from the proviso that Locke attaches to his labor theory of property (in the *Second Treatise*), where persons are allowed to appropriate natural resources (as persons now appropriate atmospheric absorptive capacity) only insofar as this does not prevent others from doing the same—or, as Locke writes, so long as persons leave "enough, and as good" for others. The basic idea is that where resources are finite, the appropriation by some of a scarce resource has the effect of harming others, because their opportunities for future appropriation are thereby diminished. Given the finite and increasingly scarce capacity of the atmosphere to absorb GHGs, the overappropriation of the atmosphere by some countries violates this Lockean proviso, leaving too little atmospheric space for others, and in effect preventing other countries from being able to develop, because insufficient atmospheric space remains to accommodate the additional growth in developing-country per capita emissions endemic to such growth as the result of the high emission rates in industrialized nations. Although Locke does not advocate the equal allocation of natural resources under these circumstances, he does recognize a key limit on appropriation, and on claims based in rights of property or prior use. Opponents of GHG regulation efforts that ground their arguments for a "minimal state" in the Lockean theory of property should realize that the central intellectual figure in the libertarian tradition long ago laid the groundwork for an international regulatory regime designed to allocate national emission shares in a manner that avoids the problem Locke identifies. If nothing else, the Lockean proviso stresses that the appropriation of natural resources must be subject to a principle of distributive justice, not left (as claimed by some contemporary neo-Lockeans) to individual choice or laissez-faire nonregulation.

To go from justified limits on appropriation to more equitable per capita emission shares requires several additional steps, however, and relies on the idea of cosmopolitan justice, in which egalitarian theories of distributive justice apply across national borders. Charles Beitz's resource redistribution principle,[20] for example, argues for the egalitarian (or equal, if also subject to something like Rawls's difference principle) allocation of natural resources, based on the logic of the Rawlsian thought experiment of the original position combined with the observation that the de facto global distribution of natural resources is morally arbitrary. Even if we reject Beitz's plausible contention that this principle applies to the allocation of *all* natural resources including those now geographically located within national borders and so subject to legal rights of ownership, and apply egalitarian distributive principles only to atmospheric absorptive capacity, which uniquely transcends national borders and so has no prior ownership claims—the weakest possible interpretation of Beitz's argument for cosmopolitan justice yet one seemingly immune to its standard criticism—his analysis makes a strong case for the equitable per capita allocation of emission shares, subject only to several side constraints concerning population growth.[21] Given equal per capita emission shares, no nation would be allowed any greater capacity to industrialize than any other, nor would any be granted the license to consume more than any other (which can inhibit social as well as economic development), so such an allocation of national emission shares would amount to the guarantee of an equal right to develop, at least so long as the absorptive capacity of the atmosphere can accommodate *any* development.

How does a right to develop emerge from the analysis of atmospheric absorptive capacity as a shared resource, combined with Beitz's resource redistribution principle? Insofar as residents of developing nations are understood to have a right to develop, there is not only a correlative negative duty on the part of the world's affluent nations to refrain from interfering with that development, but also a positive one to provide certain kinds of assistance in order to facilitate it. A right to develop, in other words, correlates with both positive and negative duties, and the recognition of such a right requires at minimum that nations not be *prevented* from realizing the benefits of development (as would be the case with overly restrictive emission caps), requiring industrialized nations to yield some of the atmospheric space that they now claim through their much larger per capita emissions in order to accommodate the GHG emission growth that accompanies development. These are minimal correlative duties—more extensive duties of assistance in the form of sustainable development aid are likely required as part of the egalitarian implications of cosmopolitan justice as applied to a global climate regime.

Due to the nonexistence of any sort of formal limits on GHG emissions during the period in which the world's affluent nations developed economically, their processes of industrialization were uninhibited in ways that can no longer be allowed for nations that have yet to undergo such processes of transformation. Should the industrialized nations now impose emission caps on developing countries such that their ability to industrialize was inhibited, they would in effect be prohibiting their further development, or interfering with their right to develop. To accommodate their interest in development, sufficient atmospheric space must be freed up under a global emission cap to allow for GHG emission increases by developing countries, and this space can only come at the expense of decreased emissions from industrialized nations. As Baer argues, some practical equivalent to a recognized right to develop, accompanied by corresponding decreases in industrialized-country emissions, may be necessary if developing countries like Brazil, India, and China are to be voluntarily brought into a system of binding emission caps, noting: "Everyone in the developing world cannot emit at the high rates of the North, but why should developing countries agree to restrictions that bind them to their current, much lower per capita rates or that restrict their economic growth?"[22] As previously observed, they cannot voluntarily agree to such restrictions, so the climate regime must be fair (to them) if it is to be effective (in including them).

If a meaningful right to develop is to be recognized and advanced by a global climate regime, it must avoid structuring incentives in such a way that developing countries are encouraged to sacrifice long-term economic viability for short-term gain. Rather, it must promote long-term human and economic development goals that are environmentally as well as socially and economically sustainable, and do so in a manner consistent with the climate convention's expressed goals of promoting equity and responsibility. This concern for equity and responsibility should not be dismissed merely as a secondary commitment to the primary goal of avoiding catastrophic climate change, for the environmental problem of anthropogenic climate change is also a problem of global justice, and so cannot be remedied unless the international response to it aims to promote justice itself while limiting GHG emissions. The right to develop, in other words, is a right grounded in ideals of justice, which seek to guarantee that the "natural lottery" of birth not continue to dictate the radically unequal life prospects that currently attach to one's nation of residence. Global climate may be only part of the complex causal chain that produces this unjust inequality, but resource exploitation patterns that contribute to climate change also lead to global inequality, and the predicted effects of climate change include the imposition of negative externality effects that significantly exacerbate that inequality. Given the interrelation between global inequality and environmental degradation—as the Brundtland Report aptly notes, global inequality is a primary cause of stress on environmental resources, and environmental degradation is primary cause of that inequality—a right to develop (within sustainable limits) is grounded in the nature of anthropogenic climate change itself. Both global justice and climate change must be addressed at once, and as manifestations of the same set of problems.

Conclusions

Insofar as persons have an interest in flourishing, and are not limited in their interests to mere survival, then they must also be recognized (as Hayward argues) as having a basic interest in an adequate environment as well as a less basic one to development. Since rights exist in order to protect interests, a strong case can be made from the critical importance to human welfare of climatic stability for a right to an adequate environment with the corollary that the right includes climatic stability. Such a right entails duties to ensure that this claim is met as well as a system of compensation when it is not. Given this interest in human flourishing, we can also posit that persons have other rights as well that guard against current and future threats to or constraints on that interest, including a right (with positive and negative dimensions) to human and economic development. While such a right cannot be unlimited—a right to development, for example, does not entail permission to deplete resources or befoul the environment—it must trump rights or other claims that are less basic to human flourishing, including those implicitly made by or on behalf of residents of industrialized nations whose selfish desire to continue producing excessive luxury emissions is mistakenly taken to justify placing overly restrictive GHG emission limits on poor countries that effectively prevent their further development. As Shue argues, more basic interests outweigh less basic ones, so more basic rights must also trump less basic ones. The right to develop cannot trump the right to survival emissions, nor can it trump the equally basic right to an adequate environment. But the former must be recognized as making a more compelling claim to limited atmospheric space than do those de facto claims now being made on that space by the relatively affluent residents of industrialized nations. These individuals selfishly seek to protect and enlarge their undeserved (following Beitz) advantages by denying to the less advantaged a prerogative (sufficient atmospheric space to accommodate development) on which their present prosperity is largely based.

The right to develop—with its positive duty of assistance as well as its negative duty not to impose practical or legal constraints on development—can be violated by a climate regime that places excessively restrictive caps on the emission limits assigned to developing nations not currently assigned binding caps under the Kyoto Protocol. But it also weighs against the right to climatic stability; indeed, many see these two rights claims as existing in fundamental opposition. Since the right to development is less basic than either of the other two rights discussed above, because the interest in human flourishing is less basic than that in mere survival, it must be limited by those two more basic rights. Nations cannot be entitled (as a matter of right) to increase their luxury emissions to that point where dangerous climatic instability threatens the very existence of current and future peoples and persons, nor can such development be financed by the denial to any of some basic level of survival emissions. But the interests in (sustainable) development can nonetheless cogently be formulated in terms of rights, and the luxury-emission budgets of developing countries may defensibly be financed by cuts in the luxury-emission budgets of industrialized ones. Since luxury emissions connote liability for the harm associated with climate change, without which the uncompensated harm of anthropogenic climate change would constitute another sort of global injustice, it is incumbent upon any fair and effective global climate regime to minimize the need for such liability, because it violates basic rights and so trumps any development interests of which luxury emissions are a component.

If all persons (including future generations) have a basic right to climatic stability, which may be violated by anthropogenic climate change, then aggregate global GHG emissions must be capped at a level at or below the atmosphere's capacity to absorb those emissions. If all have a basic right to survival emissions, then the costs associated with achieving those necessary reductions from current emissions must be assigned on the basis of historical luxury and not survival emissions, because the latter cannot serve as the basis for liability, and must grant nations and persons entitlement (to which no future liability attaches) to some basic minimum per capita level of emissions. Recognizing a less basic right to develop along with these two basic rights requires that developing countries be allowed per capita emission shares that include both survival and luxury emissions—with the latter being a necessary but insufficient condition for development—and that the interest served by this right trumps less basic interests against which it often competes, including those in private property and national sovereignty. If all three of these rights are to be recognized at once, and brought into a stable balance, then per capita emission shares assigned to industrialized nations must be significantly reduced from present levels as well as those mandated under the Kyoto Protocol, which likewise unfairly freeze the wide global economic inequalities reflected in its 1990 baseline formula. This approach will not only meet the UNFCCC's mandate of avoiding dangerous anthropogenic interference with the planet's climate system, but will also accommodate the development interests of residents of poor nations. In the interest of promoting global equity while at the same time protecting these three kinds of environmental rights, the allocation of luxury emissions among the world's nations and persons ought to be far less unequal than it is at present, and may need to contract and converge on nearly equal per capita shares.

Notes

A different version of the argument in this chapter appears in Steve Vanderheiden, *Atmospheric Justice: A Political Theory of Climate Change* (New York: Oxford University Press, 2008), portions of which are used here by permission.

1. United Nations, *United Nations Framework Convention on Climate Change* (New York: United Nations, 1992).

2. According to the Intergovernmental Panel on Climate Change's *Third Assessment Report* (Geneva: Intergovernmental Panel on Climate Change, 2001), "The impacts of climate change will fall disproportionately upon developing countries and the poor persons within all countries, and thereby exacerbate inequities in health status and access to adequate food, clean water, and other resources." An ineffective climate regime would allow this inequitable pattern of letting industrialized nations transfer the costs of their affluence to poor countries.

3. While emission caps might be allocated at either the multinational (as in the case of the European Union, which has a "bubble" cap encompassing all its member states) or subnational levels (perhaps to specific groups or regions within a nation, if not to individual persons), such allocation schemes still presuppose that caps are also awarded at the national level. In the former case, the European Union has allocated its aggregate cap among member nations, and in the latter case any group-based emission shares would necessarily need to be drawn from relevant national emission shares. I refer to national emission shares, then, to reflect the practical necessity of considering national emissions along with any other group-based shares.

4. "The Leader of the Most Polluting Country in the World Claims Global Warming Treaty Is 'Unfair' Because It Excludes India and China" (Centre for Science and Environment, 2001, www.cseindia.org/html/au/au4_20010317.htm; italics in original).

5. By "climate convention" I mean the various treaties and international agreements (including but not limited to the 1997 Kyoto Protocol) developed under the auspices of the UNFCCC.

6. Tim Hayward, *Constitutional Environmental Rights* (New York: Oxford University Press, 2005), 7.

7. Hayward, *Constitutional Environmental Rights*, 11.

8. Hayward, *Constitutional Environmental Rights*, 47–48.

9. Declaration of the United Nations Conference on the Human Environment (Stockholm: United Nations Environment Programme, 1972).

10. Its aims of protecting human dignity and welfare are explicitly anthropocentric, so it cannot connote duties in which nonhuman animals or ecosystems are the exclusive beneficiaries.

11. Brian Barry, "Sustainability and Intergenerational Justice," in Andrew Dobson, ed., *Fairness and Futurity*, 93–117 (New York: Oxford University Press, 1999), (quote on 97).

12. Henry Shue, *Basic Rights* (Princeton, NJ: Princeton University Press, 1980), 18–19.

13. Shue, *Basic Rights*, 23.

14. Shue, *Basic Rights*, 34.

15. Shue, *Basic Rights*, 126–127.

16. Henry Shue, "Climate," in Dale Jamieson, ed., *A Companion to Environmental Philosophy* (Malden, MA: Blackwell, 2001): 449–459 (quote on 451).

17. Shue, "Climate," 454.

18. Thomas Athanasiou and Paul Baer, *Dead Heat: Global Justice and Global Warming* (New York: Seven Stories Press, 2002), 75.

19. Athanasiou and Baer, *Dead Heat*, 128.

20. Charles R. Beitz, *Political Theory and International Relations* (Princeton, NJ: Princeton University Press, 1979), especially 138–149.

21. For example, nations might be assigned equal per capita emission shares in the present, but in the future be assigned greater or lesser shares based on their (presumably voluntary) rates of population growth.

22. Paul Baer, "Equity, Greenhouse Gas Emissions, and Global Common Resources," in Stephen H. Schneider, Armin Rosencranz, and John O. Niles, eds., *Climate Change Policy: A Survey*, 393–408 (Washington, DC: Island Press, 2002) (quote on 394).

4

Environmental (In)justice in Climate Change

Martin J. Adamian

The prospect of global climate change provides an excellent case from which to assess the adequacy of theoretical foundations of international law and the ways our understanding of law and justice shape the use of legal pronouncements to address global environmental problems. This chapter examines the global climate regime,[1] the idea of international environmental justice, and Critical Legal Studies (CLS) in an attempt to assess the possibility of advancing justice in the international context of global climate change. We know that human activities have had a significant impact on the global climate system, and if current trends continue, these will lead to potentially disastrous effects on the environmental, social, and economic interests of all states and will have profound consequences for every aspect of human societies. No state can hope to arrest climate change on its own, but collective action undertaken by sovereign states with different socioeconomic and environmental circumstances would be extremely difficult. In addition to the challenges it poses to politics, society, and the environment, climate change challenges mainstream legal theory and practice as well as its foundational normative concepts of state responsibility, sovereign equality, and the centrality of states within the international legal system.

Predictions of global climate change raise a number of very difficult practical, moral, and ethical issues that states have only begun to address. Currently, international environmental law is seen as the main mechanism for addressing the anthropogenic causes and effects of climate change, along with their associated social problems and the ethical issues they generate. In this chapter, I examine the development and application of international law in a climate change mitigation regime, starting with an examination of international environmental law generally, considering how international environmental justice is conceptualized, and then applying lessons from CLS in order to assess the prospects of using these legal tools to promote justice.

International Environmental Law

International environmental law is arguably the most dynamic area of international law, and is described as having "had the greatest impact, ultimately constituting a powerful factor pushing toward a transformation of the fundamental basis of international law."[2] Yet a cohesive body of international law establishing environmental regulation among sovereign states remains relatively undeveloped. Historically, environmental problems have largely been considered matters of domestic concern within the sovereign jurisdiction of each state, and thought to be properly addressed by domestic regulation rather than international law. Within the domain of international law, justice has meant legality, sovereignty, equality, and fairness of treatment among states, but conventional international law was not well equipped to deal with the variety of issues that have arisen regarding international environmental justice.

Over time, states began to accept, on a voluntary and reciprocal basis, a variety of legal restrictions in order to pursue their shared objectives. This became increasingly true as the international order as well as each state's environmental integrity became more and more interdependent. Nevertheless, the sovereign equality of states and the voluntary acceptance of international obligations remain fundamental to modern conceptions of international law. States have established a number of international organizations and vested them with limited legal powers sufficient to achieve particular common goals, but there remains no central legislative body at the international level. As a result, existing international environmental rules are "a patchwork, reflecting a piecemeal, fragmented and *ad hoc* response to problems as they have emerged."[3] However, international law is increasingly used to resolve disputes between international actors, particularly those between states. Whether it is up to the task remains very much in dispute.

International environmental law has developed considerably since the 1972 United Nations Conference on the Human Environment, held in Stockholm, which saw the first major acknowledgment by the industrialized nations of the importance of multilateral efforts to deal with transboundary environmental problems.[4] By the time of the 1992 UN Conference on Environment and Development (popularly known as the Earth Summit), held in Rio de Janeiro, it had become clear that environmental concerns would occupy a central place on the world political agenda, spurring the creation of what some now see as a fledgling system of global environmental governance. According to the World Resource Institute, this system consists of three elements:

1) international organizations such as UNEP, the United Nations Development Programme, the Commission on Sustainable Development, the World Meteorological Organization, and dozens of specific treaty organizations.

2) a framework of international environmental law based on several hundred multilateral treaties and agreements.

3) financing institutions and mechanisms to carry out treaty commitments and build capacity in developing countries, including the World Bank and Specialized lending agencies such as the Multilateral Fund and the Global Environment Facility.[5]

In total, there are now hundreds of bilateral and regional treaties as well as organizations that deal with transboundary and shared resource issues, and more than 900 international agreements with at least some environmental provisions.[6]

As decision-making authority gravitates from the national to the international level, the question of the legitimacy of international governance has begun to receive more attention. While its sources and scope of power are identifiable, the legitimacy of this international governance system is far more difficult to ascertain.[7] Until recently, international institutions have generally been relatively weak and have exercised so little authority that the issue of their legitimacy has barely arisen, with scholars more often focusing on the causal role of international institutions than on their legitimacy. To the extent that one can speak of global governance, its authority has rested on the consent of the very states to which it applies. Modern theories of legitimacy often attempt to base the legitimacy of governmental authority on the consent of the governed, but as Daniel Bodansky notes: "[In] international law, the strongly consensualist basis of obligation has tended to moot the issue of legitimacy."[8] As international institutions have gained greater power and authority, however, calls for more democratic participation and consent in international environmental law have begun to be voiced, but as Bodansky notes,

Democracy can mean different things—popular democracy, representative democracy, pluralist democracy, or deliberative democracy to name a few. What might it mean in the context of international environmental law? Democracy among states or among people? A system of majority decision making or simply greater participation and accountability? And if the latter, participation by whom and accountability to whom? Abraham Lincoln once characterized democracy as government "of the people, by the people, and for the people." But who are "the people" in this connection?[9]

Despite making important strides, the current international governance system remains weak and ineffective. Without a centralized government or sovereign political authority to oversee global governance, international agencies often duplicate some efforts while collectively failing to address other issues. Moreover, these organizations are forced to rely on individual states to carry out their policies, but states are reluctant to relinquish their sovereignty and prerogative to pursue their own national interest. As will be further explored below, CLS shows how international environmental law and governance may perpetuate domination and subordination without effectively addressing the problems they were created to tackle, a tendency illustrated by its application to the problem of global climate change.

Climate Change

Two major international climate treaties were negotiated in less than a decade: the UN Framework Convention on Climate Change (UNFCCC) in 1992 and the 1997 Kyoto Protocol,[10] and both have been significantly elaborated through additional legal instruments and decisions adopted under the rubric of the UNFCCC. The international legal and institutional framework established by these instruments, along with their relationship to other international issues, is as complicated and far reaching as the climate problem itself, with the underlying complexity of the climate problem and the sheer pace of scientific and political developments as contributing factors. The protocol, which is one of the most innovative and ambitious international agreements ever negotiated, sets greenhouse gas emission targets for all industrialized countries, with an overall cut of 5 percent from 1990 baseline emission levels but ranging from cuts of 8 percent for some countries to increases of 10 percent for others.[11] Emission targets also cover certain carbon sequestration activities in land use, land-use change, and the forestry sector, based on specific rules, and must be met by the commitment period of 2008–2012.[12] The protocol also utilizes flexibility mechanisms in determining compliance, including joint implementation, clean development mechanisms, and emission trading to help states meet their targets.

Throughout the negotiations leading up to and following the Kyoto Protocol, divisions emerged between industrialized and developing countries.[13] Most developing countries wanted to focus on implementation of existing commitments declared by the UNFCCC, while industrialized countries were interested in launching a post-Kyoto round covering developing countries. The stakes were raised by the fact that the U.S. Senate had made the "meaningful participation" of developing countries a condition for its ratification of the protocol.[14] In March 2001, newly elected President George W. Bush explicitly rejected the protocol, standing virtually alone in opposing specific targets and timetables for stabilizing CO_2 emissions, and continued to emphasize scientific uncertainties involved in forecasts of global warming and also expressed concern about the economic impacts of CO_2 stabilization policies.[15] Nevertheless, multilateralism, backed by scientific consensus in the form of the 2001 *Third Assessment Report*, issued by the Intergovernmental Panel on Climate Change (IPCC), held the climate change regime together. Following its 2004 ratification by the Russian government, the protocol was set to enter into force, and at this point the climate regime is primarily in an implementation phase, with Parties focused on putting into practice the large body of rules that are now in place to guide their efforts to combat climate change, and implementing these rules and regulations through national legislation.

The divisions between industrialized countries of the North and the developing countries of the South demonstrate the significance of power asymmetries in global environmental politics, and highlight the need to employ a critical conceptual lens such as that provided by CLS in order to more clearly assess the prospect of achieving international environmental justice within the context of climate change. According to the CLS critique, traditional liberal legal theory does not fully address the various questions of justice that must be taken into consideration if we are to effectively address issues such as global climate change while pursuing international environmental justice. But before we can appreciate the force of this critique, we must first ask: What is international environmental justice?

International Environmental Justice

It has recently become clear that the causes and consequences of global environmental degradation cannot be addressed without also tackling inequality and injustice.[16] Issues of justice as they relate to environmental degradation have most often been approached domestically, with the term *environmental justice* emerging from the growing recognition that people of color and those with low incomes, far more often than other segments of the population, live and work in areas where environmental risks are high.[17] Such communities are disproportionately likely to become unwilling recipients of hazardous waste sites, incinerators, and industrial production facilities, and their residents more likely to be exposed to pesticides and radiation, underscoring the sharp inequality of opportunities to breathe clean air, drink clean water, enjoy pristine wilderness, or work in a clean, safe environment, both in the United States and internationally.

Particularly on the international level, environmental justice remains a contested concept; not everyone agrees about how to define justice, nor do they agree about how to weigh competing claims.[18] The UN Charter refers to justice but does not define it,[19] noting instead the aim "to establish conditions under which justice and respect for the obligations arising from treaties and other sources of international law can be maintained."[20] Furthermore, the charter provides that "all members shall settle their international disputes by peaceful means in such a manner that international peace and security, and justice, are not endangered."[21] Exactly how these provisions help reconcile competing notions of justice is unclear.

Decisions about environmental policy, like so many other policy decisions, often result in clear winners and losers. Inherent in the notion of environmental justice is the more general concept of justice itself; indeed, some claim that environmental justice is concerned more with the topic of justice than the environment.[22] To properly understand environmental justice, some common understanding of justice and how it can be pursued on the international level are needed. In international law, compliance may have more to do with the concept of fairness than with dispute settlement mechanisms or sanctioning regimes. As noted in one study on compliance, "People obey the law because they believe that it is proper to do so, they react to their experiences by evaluating their justice or injustice, and in evaluating the justice of their experiences they consider factors unrelated to outcome, such as whether they have had a chance to state their case and been treated with dignity and respect."[23] Here, the concept of fairness involves two features. First, it concerns the ex ante affirmation of a political order. The degree to which a new law or judicial opinion is likely to be perceived as fair will thus depend in part on the extent to which it has been formulated by a discursive process where those most likely to be affected have been invited to present their views, and all the discourse's participants accept a need for mutual accommodation without reserving any matter as nonnegotiable.[24] Second, the concept of fairness concerns the ex post affirmation of the decisions that emanate from a political order.

The fairness of international law, as of any other legal system, may be judged, first, by the degree to which its rules satisfy the participants' expectations of a justifiable distribution of costs and benefits, and, second, by the extent to which the rules are made and applied in accordance with what the participants perceive as the right process.[25] Both aspects of fairness are essential to a successful legal order. The importance of fairness in governance, both ex ante and ex post, has been recognized by numerous scholars, from Immanuel Kant to Jürgen Habermas.[26] John Rawls acknowledges the primacy of justice as fairness, claiming that "justice is the first virtue of social institutions, as truth is of systems of thought. A theory, however elegant and economical, must be rejected or revised if it is untrue; likewise laws and institutions no matter how efficient and well-arranged must be reformed or abolished if they are unjust."[27]

However, relatively little attention has been paid to exactly what the "justice" of "environmental justice" refers to, particularly in the realm of social movement demands. Most understandings of environmental justice refer to issues of equity or the distribution of environmental harms and benefits, but defining environmental justice in terms of equity is incomplete, because activists, communities, and nongovernmental organizations call for much more than just distribution. As William Aceves suggests, participation and recognition are also essential to ensuring fairness in the international system. Similarly, David Schlosberg suggests that the justice demanded by global environmental justice is really threefold: it consists in equity in the distribution of environmental risk, recognition of the diversity of the participants and experiences in affected communities, and participation in the political processes that create and manage environmental policy.[28] Most theories of environmental justice are incomplete theoretically, because they continue to solely emphasize distributive justice and so fail to adequately integrate the related realms of recognition and political participation. Furthermore, such theories are insufficient in practice, since they are disconnected from the more thorough and integrated demands and expressions of many social movements for environmental justice globally. Schlosberg's central argument is that a thorough notion of global environmental justice needs to be locally grounded, theoretically broad, and plural—encompassing issues of recognition, distribution, and participation.

Similarly, Iris Young criticizes liberal theories of justice like that of Rawls for failing to adequately recognize the full scope and context of justice, arguing that while theories of distributive justice offer models and procedures by which distribution may be improved, none thoroughly examines the full social, cultural, symbolic, and institutional conditions underlying unjust distributions. Moreover, she claims, such theories tend to assume the goods in question to be static, rather than the dynamic outcome of various social and institutional relations, and so mistake the nature of the problem from which the maldistribution of goods is only one manifestation. "Distributional issues are crucial to a satisfactory conclusion of justice," she writes, but "it is a mistake to reduce social justice to distribution."[29] Injustice, Young claims, cannot be based solely on inequitable distribution, because there are reasons why some people get more than others. Part of the problem of injustice and cause of unjust distribution, she argues, is a lack of recognition of group difference. If social differences are attached to both privilege and oppression, Young contends, justice requires an examination of those differences to undermine their distributive effects. Recognition is essential, since its absence, demonstrated by various forms of insults, degradation, and devaluation at both the individual and cultural level, inflicts damage on both oppressed communities and the image of those communities. Likewise, she claims, the lack of recognition is an injustice not only because it constrains people and does them harm, but also because it is the foundation for distributive injustice.

Most scholars working on justice theory start with a liberal focus on distributive justice, rather than recognition or participation. As Nicholas Low and Brendan Gleeson note, "The distribution of environmental quality is the core of 'environmental justice'—with emphasis on distribution."[30] While Low and Gleeson advocate greater political participation as a means toward environmental justice, drawing causal links between participation, inclusive procedures, and public discourse oriented toward environmental justice, these concerns are not incorporated into their ideal principles or practices of ecological justice. Their focus is on global, cosmopolitan institutions rather than those at the local, community level, and while they acknowledge the contextual and cultural bases of the meanings of both the terms *environment* and *justice*,[31] they cannot bring this notion of cultural difference into their definition of either environmental or ecological justice.

In sum, a broader understanding of international environmental justice is needed in order to discern what justice might look like in the context of global climate change, as well as to assess the possibility of achieving international environmental justice through the use of international environmental law. CLS analysis can be an essential component of such an approach, since it broadens conventional liberal conceptions of justice and so assists in better understanding international environmental justice. The insights it offers are indispensable in grasping international environmental problems like climate change.

The CLS Critique of International Environmental Law

CLS includes a heterogeneous body of legal theory that borrows from several theoretical traditions in developing a critique of the liberal legal tradition. According to this critique, liberal legal theory, as a product of mainstream legal studies, contributes to the current predicament in international politics by fostering the belief that international law can be relied on to achieve global justice, particularly in the case of global environmental degradation. As suggested above, however, conventional liberal theories of law and justice are conceptually inadequate, because they overlook several key dimensions of justice, so alternative approaches like CLS are needed in order to critically assess the foundations of international environmental law.

Critical legal scholars assert that behind all legal doctrine and legal systems stand political judgments that reflect the unarticulated domination of the makers and shapers of the law. As Robert Gordon summarizes this critique, it holds that "supposedly universal norms are deployed for the benefit of a particular class."[32] Its proponents maintain that the logic and structure attributed to the law grow out of power relationships within society, that the law exists to support the interests of the party or class that forms it, and that it is merely a collection of beliefs and prejudices that legitimize the injustices of society. According to this analysis, the wealthy and powerful use the law as an instrument for oppression in order to maintain their place in the hierarchy. The basic idea is that the law is inherently political, favoring some interests over others, rather than being (as liberalism maintains it should be) neutral or value free. Some of the lessons from CLS, as well as some practical considerations about the unique nature of the international legal system that arise from it, allow us to assess the capacity of international environmental law to promote justice in the context of climate change, as the remainder of this section aims to demonstrate.

Individual Subjective Values

A fundamental assumption of classical liberalism as well as liberal theories of law is the focus on individual subjective values. Liberals assume that the interests of distinct individuals are internally consistent and conducive to the general good, yet liberalism fails to properly account for the unique interests of collective entities through an adequate accounting of objective value. This is particularly problematic when we attempt to reconcile the interests of those in Western industrialized democracies that explicitly recognize and pursue individual liberty with those residing in developing countries that struggle to fit their own notion of communal values into an international system dominated by liberal notions of individualism. Many developing countries are the home to indigenous groups that fight for collective rights of self-determination, but within mainstream liberalism only individuals are recognized as having rights. While some liberals such as Will Kymlicka urge the recognition of group as well as individual rights, the very notion of rights, and especially individual rights, is unknown in many African languages and is of limited significance in traditional Islamic, Hindu, and Buddhist communities.[33] If there is to be any meaningful protection of personal interests in such nonliberal contexts, the communities in which these persons reside must be protected. After all, CLS pointedly asks, how do we understand the individual if not by reference to the community from which they develop their sense of self? Liberal theories of law, in failing to adequately account for communal values, treat communal values merely as the sum of the individual values that make up that community, thereby failing to account for group-based interests that are more than the sum of the group's individual parts.

In applying international environmental law to problems of climate change, duties and obligations are typically outlined in reference to states based on their level of development. Many developing states are home to the indigenous groups noted above, providing an additional point of contention when international law fails to recognize these groups and their communal values. In addition, there is no way to ensure that these groups or other subnational units are able to participate in or have their interests heard by an international climate regime, given their minority status within nations that send delegates to climate policy conferences. The result is that international law fails to adequately safeguard and represent the interests of all groups equally, and it is therefore not surprising that not all groups benefit equally from its application.

Roberto Unger points out that the principle of subjective value is closely related to the liberal conception of rules as the basis of order and freedom in society, suggesting that ends are viewed as individual in the sense that they are always the objectives of particular individuals. This is true in regard to international law as well as law in the domestic context, although on the international level we must deal with the aggregation of individual interest, making the issue more problematic when applied to international law and climate change since, as discussed above, the political doctrine of liberalism does not acknowledge communal values. The two basic ways liberalism defines the opposition of rules and values correspond to two ideas about the source of laws and to two conceptions of how freedom and order may be established. Within the liberal tradition, laws must be impersonal in order to establish freedom and order. They must represent more than the values of an individual or of a group, since rules based on the interest of a single person or classes of persons contradict the basis of freedom. As CLS scholars have argued, however, there is no guarantee that international law represents anything more than the interests of certain groups, such as the wealthy and powerful.

Domination and Subordination under a Rule of Law

Three things are necessary to support a rule of law in any given system: a lawmaking process, a process of law enforcement, and an adjudication process. In considering whether a lawmaking process exists, it must be acknowledged that although the UNEP and the United Nations more generally include most states in the international system, not all states are included. Further doubts arise about the adequacy and effectiveness of a climate regime that fails to include within its regulatory apparatus the largest contributors of global greenhouse emissions, because China is exempt from the first round of emission caps and timetables and the United States has been unwilling to sign on to the protocol. The United States has historically relied on fossil fuels to propel its economic growth, and its current per capita consumption of such greenhouse-intensive energy far exceeds that of the most of the rest of the world. China has the world's largest population and among the fastest-growing per capita emissions, and is expected to surpass the United States in total national emissions sometime in this century. These exclusions alone call into question the adequacy of a lawmaking process that is not binding on all nations, particularly those that are most responsible for causing climate change and so arguably should be held responsible for addressing the issue.

Also worth noting is the fact that although the UN system has been utilized to coordinate international action on a variety of issues, it is not a full world government. The UN Charter explicitly recognizes state sovereignty as a fundamental limitation on its power, and it is highly unlikely that the UN would have such broad support without such safe-guards, which protect individual states from the monopoly of power that might be used to pursue collective goals that are not also within the national interests of each state. Superpowers like the United States had little incentive to join into such an organization without the ability to control its agenda, so the UN Security Council was established to reflect the power hierarchies that existed at the time. Although the UN is an international legislative institution in which rules and regulations are debated and created, CLS critics might question the ability of the organization to adequately address issues that adversely affect the least powerful groups in the international system, given the way power is inequitably allocated within the institution.

As a hegemonic power, the United States has considerable influence and a formal veto power that permits it to control which issues are addressed and how, while less powerful states are unable to effectively utilize UN power on their behalf unless they can convince the major players that it is also in their best interest to do so. Hence, issues like climate change cannot be adequately addressed insofar as this requires countries like the United States to make sacrifices that could possibly adversely affect their economy. The efficacy of international law is chal-lenged by the traditional realist claim that states will only follow inter-national law when it is in their national interest. While the United States was successful in incorporating flexible mechanisms that would seem to allow some progress toward addressing these issues, such mechanisms arguably displace pollution without adequately minimizing the effects of global greenhouse gas emissions in a way that will solve the practical and ethical problems that surround the possibility of global climate change. Ultimately, even with the inclusion of flexible mechanisms, the United States did not believe that the Kyoto Protocol was consistent with its national interest.[34]

In sum, some international lawmaking process may exist, but serious questions remain concerning the ability of such a process to produce fair and equitable rules and regulations that are binding on the most significant international actors. Furthermore, and as a result of the inadequacies of the system itself, the rules and regulations that have thus far been established fall short of what is required to sufficiently reduce and offset global greenhouse gas emissions in order to avert some dire predicted consequences of climate change. Even the specific commitments of the Kyoto Protocol will, according to critics, neither "halt global emissions growth, nor have a discernible impact on economic growth,"[35] warranting this sort of criticism of international environmental law insofar as it only gives lip service to the problems it is meant to deal with.

Even if a lawmaking process exists, also necessary for a rule of law are adequate law enforcement and adjudicatory processes. With regard to law enforcement, the lack of an effective world government makes it extremely difficult to ensure compliance with rules and regulations in an international system that revolves around the sovereignty of states, ultimately resulting in the decentralized nature of the international system. Since much of international law is nonbinding, and since for that small portion of law that utilizes binding timetables or regulations there is no international police force capable of sufficiently monitoring and enforcing these pronouncements, the existence of adequate executive power to enforce international environmental laws like those wielded by the global climate regime likewise remains seriously in doubt.

Adjudication is equally problematic. Even when satisfactory processes have been established to monitor and enforce compliance with international law, also required is some sort of adjudicatory body for resolving disputes about noncompliance and meeting the overall objectives of the regime. The use of name-and-shame polices can only go so far when attempting to reconcile the competing interests of the world's economic and military superpowers, particularly those that have a tendency to follow international law only when it is consistent with their economic and military objectives.

Whether a rule of law exists is only part of the issue. Assuming that international environmental law is indeed positive law, it remains to be determined whether it fulfills the promises that liberal legal theory suggests. The common solution to the problems of order and freedom within the liberal tradition lies in the formulation and application of impersonal rules or laws. This is problematic in the climate regime for all of the same reasons discussed above, because existing international lawmaking and executive processes are simply inadequate to ensure that the rules, laws, and norms that are established are impersonal. Instead, they tend to reinforce the global power hierarchies that already exist and form the basis for the creation of the international institutions within which these laws are debated and created.

Critical legal scholars suggest that the rule of law is a mask that gives existing social structures the appearance of legitimacy and inevitability. Hence, they suggest, the liberal claim to neutrality is pretextual and thus conceals unacknowledged interests and relationships of power. Within the liberal tradition, law is viewed as an indispensable mechanism for regulating public and private power in a way that prevents oppression and domination, but the liberal reliance upon the rule of law as well as its defense of individual liberty have historically justified a system of private property, which in turn has ultimately supported inequity on the national and international levels. In global climate policy, flexible market mechanisms have been adopted in an attempt to address issues of international inequity while incorporating a market value for the externalities of global capitalism, but this liberal reliance on a system of private property rights justifies a class-based system that benefits certain groups at the expense of others. This is done in the name of individual liberty, but CLS critics would challenge this system by asking of the application of international environmental law to global climate policy: Whose liberty is being defended? Is individual liberty protected through these flexibility mechanisms, or is the freedom of corporations to exploit and oppress being nurtured instead? Ultimately, the corporation as a legal entity benefits most from environmental valuations by allowing environmental degradation when it is economically defensible.

In sum, liberalism assumes that laws are universal, consistent, public, and capable of coercive enforcement, but it is unrealistic to assume that laws can be any of the above when they are produced in an international system that simply reinforces the power hierarchies that already exist. CLS can be utilized to help us understand how the system reinforces such hierarchies while giving the perception of legitimacy. Since the resort to a set of rules as the foundation of order and freedom is a consequence of the subjective conception of value, we must accept that international laws will not be universal or consistent. Were they universal and consistent, however, still necessary is an appropriate technique of rule application from which we can deduce conclusions from premises and choose the most efficient means to accepted ends. Liberal theories of adjudication view the task of applying law either as one of making deductions from the rules or as one of choosing the best means to advance the ends the rules themselves are designed to foster. This resort to a set of rules as the foundation of order and freedom is a consequence of liberalism's subjective conception of value.

Problem of Legislation and Adjudication

According to the CLS critique, the problem of legislation results from the fact that in order to make law one has to choose among competing individual and subjective values, and ultimately give preference to some over others. Liberal legal theory holds that a procedure for lawmaking exists on the basis of the combination of various private ends. It is believed that all individuals would subscribe to these procedures in self-interest, by which is meant the intelligent understanding of what each person needs in order to achieve their own individual and subjective goals. However, this substantive theory of freedom breaks down because of the difficulty in finding a neutral way to combine various individual subjective values.

This is true in regard to climate change as well. Already noted are the practical and ethical difficulties that plague attempts to come up with satisfactory international environmental laws that address climate change. A primary reason why the United States has been unwilling to participate in the protocol concerns issues of equity that remain in regard to the "differentiated responsibilities" and resulting timetables established for countries of the Global South. Why, ask critics of American participation in the climate regime, should the United States agree to a set of arrangements that place substantial mitigation burdens on it? While this question may be answered by noting several key differences in regard to the causes of the current climate predicament, the fact remains that it may not be in the U.S. national interest to sacrifice its standard of living and economy when the regime's obligations are not distributed equally yet its benefits are shared by all. On the other hand, given the historic global greenhouse emissions of countries like the Untied States, it is equally unreasonable to expect less developed states to assume the same responsibilities and obligations for addressing a problem that they have little responsibility for causing.

It is not enough to have a satisfactory method for rulemaking unless there is also a process for the application and adjudication of those rules. Critical legal scholars suggest that words written by someone else (i.e., the legislature) are subject to interpretation and thus manipulation by a decision maker (i.e., the judge) enforcing the written word, so the decision about how to formulate legal commands can be just as difficult as deciding how to interpret and apply those edicts. Hence, the CLS critique maintains, the problem of adjudication is really an extension of the problem of legislation; unless one interpretation of the rules can be justified over another, the liberal claim of neutrality in the adjudication of disputes must be rejected as untenable. In addition to addressing adequate procedures for making and applying laws, one must also address the standards according to which, or in what manner, the laws can be applied without violating the requirements of freedom.[36]

Critical legal scholars, such as Unger, claim that liberal legal theory is incapable of avoiding purposive legislation and adjudication, since laws are created and interpreted by individuals with no safeguards for ensuring that their decisions adequately account for the interests of all stakeholders. In this regard, CLS critics maintain that liberal theories of law are unable to reconcile concerns for legal and substantive justice, and international environmental law has the potential to suffer from the same difficulty. That is, how can we ensure that the outcomes of legal mechanisms have any connection with justice? A focus on legal justice can guarantee that the legal system treats all individuals similarly but it cannot account for the fact that individuals differ in their needs, capabilities, and responsibilities. Substantive justice, in its focus on outcomes rather than procedures, is hard to define, let alone to pursue, and having one is no guarantee of the other.

The use of international environmental law to address global climate change must contend with the problems of legislation and adjudication noted above, which largely rest on the inability of liberal legal theory to safeguard law from the domination of subjective individual values. Absent an ability to justify the supremacy of certain wills or interests over those of others, liberal legal theory has no way of achieving a satisfactory theory of legislation or adjudication. Critical legal scholars have illuminated this predicament and, in so doing, can facilitate a shift toward a focus on substantive justice.

Indeterminacy of Law

A related claim presented by critical legal scholars has been their challenge to the view that law is composed primarily of determinative rules that are logically applied by neutral adjudicators to reach predictable, correct results. The amorphous and ad hoc nature of international law makes it difficult to expect definitive results, and its use in the climate regime provides no exception to this tendency. In addition, the rules themselves are not necessarily determinative or logically applied, and it would be difficult to maintain that anything about the rules and regulations is neutral. Therefore, it is exceedingly unlikely that the presence of a rule of law (with the components noted above) would be sufficient to address a complicated issue like global climate change, with the practical and ethical issues arising in attempts to coordinate international cooperation to effectively address it.

As a result of the indeterminacy of law and legal pronouncements, international environmental law in and of itself reinforces hierarchies of power and allows countries of the Global North to maintain their privileged positions. Like law generally, international environmental law, CLS critics argue, is used to represent the interests of the wealthy and powerful, and often fails to provide freedom and order while protecting the individual liberty it seeks to embrace. Instead, it gives the appearance of legitimacy while failing to effectively address problems such as climate change. While liberal theories of law and justice alone are inadequate, CLS can illuminate the significance of law and power in a way that can usefully assist in addressing climate change while pursuing international environmental justice. Only with a broad, pluralist approach can we fully understand the theoretical foundations and limitations of international jurisprudence.

Conclusion

Three options exist for responding to global climate change: prevention, mitigation, and adaptation. Prevention is no longer an option, and adaptation is the least desirable and would be the most unjust of the three; therefore the debate is currently over strategies for mitigation. Attempts to mitigate the effects of climate change have largely taken the form of international law, but there are a number of problems with a climate regime that relies heavily on the use of international environmental law. Such a system cannot fully address the variety of practical, social, and ethical issues raised by such a complicated problem as climate change. This does not mean that international environmental law is of no value in promoting global justice and equity, but we must remain critical of the ways law is used to justify a system that benefits some at the expense of others.

Climate change challenges individuals as well as states, but while individuals may be capable of developing personal responsibility through their choices, it is unrealistic to expect states to develop global responsibility. As Prue Taylor notes,

States are not conscious entities, not living beings, but social institutions. They cannot initiate change themselves, rather they reflect and execute the change made by individuals, groups and society. Similarly, international law is not itself capable of bringing about change. International law is merely a body of treaties, principles and institutions. Its use is determined by the ability of states to incorporate international obligations within their municipal law and, at the same time, respond to new challenges.[37]

Although state action is indispensable in our effort to harness international cooperation to deal with global problems such as climate change, we must question the assumptions of an international legal system based on state sovereignty. Ultimately, sovereignty must be recreated to incorporate a concern for planetary sovereignty and the sovereignty of peoples. This will require a greater democratization of environmental governance in order to incorporate greater participation while still paying attention to and responding more effectively to local voices and local concerns rather than seeing the state as the sole arbiter of competing interests in the determination of public policy. Only then will we be able to rejuvenate the international legal system to address the practical and ethical issues that it seeks to remedy.

Notes

1. The term *regime* is used to refer to the rules, regulations and institutions relevant to a particular subject area. More specifically, in international relations, a regime has been defined as "a set of implicit or explicit principles, norms, rules and decision-making procedures around which actors' expectations converge in a given area in international relations" (S. D. Krasner, "Structural Causes and Regime Consequences: Regimes as Intervening Variables," *International Organizations* 36, no. 21 (1982): 186; Farhana Yamin and Joanna Depledge, *The International Climate Change Regime: A Guide to Rules, Institutions and Procedures* (Cambridge: Cambridge University Press 2004), 6–7). Note that this includes both binding, or hard international law, and nonbinding law, sometimes referred to as soft international law.

2. A. C. Kiss and D. Shelton, *International Environmental Law* (New York: Transnational, 1991), 2; Prue Taylor, *An Ecological Approach to International Law: Responding to Challenges of Climate Change* (London: Routledge, 1998), 3.

3. Yamin and Depledge, *The International Climate Change Regime*, 11.

4. Lorraine Elliott, *The Global Politics of the Environment* (New York: New York University Press, 1998), 7.

5. World Resources Institutes, *World Resources 2002–2004: Decisions for the Earth: Balance, Voice, and Power* (Washington, DC: World Resources Institute, 2003), 138.

6. Edith Brown Weiss, "The Emerging Structure of International Environmental Law," in Norman J. Vig and Regina S. Axelrod, eds., *The Global Environment: Institutions, Law, and Policy* (Washington, DC: CQ Press, 1999), 111.

7. Donald A. Brown, *American Heat: Ethical Problems with the United States' Response to Global Warming* (Lanham, MD: Rowman & Littlefield, 2002), vii.

8. Daniel Bodansky, "The Legitimacy of International Governance: A Coming Challenge for International Environmental Law?", *American Journal of International Law* 93, no. 3 (1999): 597.

9. Bodansky, "The Legitimacy of International Governance," 599.

10. By 1995, it had become clear that the nonbinding approaches to global warming contained in the UNFCCC were failing to make much progress. At the first COP to the UNFCCC in Berlin in 1995, the parties agreed to begin negotiations on a binding protocol on emission limitations.

11. There is no ethically defensible reason why emission quotas should be based on the state of energy consumption in 1990. See Robin Attfield, *The Ethics of the Global Environment* (West Lafayette, IN: Purdue University Press, 1999), 93. Several other bases have been put forth. Steven Luper-Foy suggests that all natural resources should be regarded as available to the whole of humanity, present and future, and should be shared accordingly (see Steven Luper-Foy, "Justice and Natural Resources," *Environmental Values* 1, no. 1 (spring 1992): 47–64). Michael Grubb proposes recognition of the equal entitlement of all human beings to access to the absorptive capacities of the planet. This view would justify per capita emissions for all countries (see Michael Grubb, *The Greenhouse Effect: Negotiating Targets* (London: Royal Institute of International Affairs, 1989), as well as his *Energy Policies and the Greenhouse Effect* (Aldershot, UK: Gower, 1990)). Shue argues that whatever quotas are set for emissions, equity requires that provisions must be made for emission entitlements that facilitate the satisfaction of everyone's basic needs (Henry Shue, "Equity in an International Agreement on Climate Change," paper presented at the IPCC workshop on "Equity and Social Considerations Related to Climate Change," Nairobi, 1994, 7–14; quoted in Attfield, *The Ethics of the Global Environment*, 93).

12. Yamin and Depledge, *The International Climate Change Regime*, 25.

13. As early as the 1972 Stockholm Conference, industrialized and developing states agreed to address the environment as well as development, but developing countries feared restrictions on their economic growth and had to threaten noncooperation and appeal to socially shared norms of social justice in order to achieve this outcome. As a result, the resolution remained vague on operational details and virtually no action was taken for almost two decades. See Peter Haas, Marc Levy, and Ted Parson, "Appraising the Earth Summit: How Should We Judge UNCED's Success?", *Environment* 34, no. 8 (1992): 6–11, 26–33. As a result of the possibility of a North-South standoff and Southern opportunism, architects of the Stockholm Declaration designed a "Resolution on Institutional and Financial Arrangements" and included an "Environment Fund" to assist

developing states in their efforts toward sustainability (Bradley C. Parks and J. Timmons Roberts, "Environmental and Ecological Justice," in Michele M. Betsill, Kathryn Hochstetler, and Dimitris Stevis, eds., *Palgrave Advances in International Environmental Politics* (Basingstoke, UK: Palgrave Macmillan, 2006), 330).

14. Byrd-Hagel Resolution, Senate Resolution 98, adopted July 1997; cited in Yamin and Depledge, *The International Climate Change Regime*, 26.

15. There are many uncertainties concerning anthropogenic climate change. Nevertheless, many suggest that we cannot wait until all of the facts are in before we respond. Dale Jamieson, for one, notes that there are uncertainties regarding the impacts of climate change as well as regarding human responses to these changes: "One thing is certain: The impacts will not be homogenous. Some areas will become warmer, some will probably become colder, and overall variability is likely to increase. Precipitation patterns will also change, and there is much less confidence in the projections about precipitation than in those about temperature. These uncertainties about the regional effects make estimates of the economic consequences of climate change radically uncertain" (Dale Jamieson, "Ethics, Public Policy, and Global Warming," *Science, Technology, & Human Values* 17, no. 2 (1992): 145). Jamieson notes further that climate change "will affect a wide range of social, economic, and political activities. Changes in these sectors will affect emissions of 'greenhouse gases,' which will in turn affect climate, and around we go again" (p. 145).

16. Parks and Roberts, "Environmental and Ecological Justice," 329.

17. See, for example, P. Mohai and B. Bryant, "Environmental Racism: Reviewing the Evidence," in B. Bryant and P. Mohai, eds., *Race and the Incidence of Environmental Hazards: A Time for Discourse* (Boulder, CO: Westview Press, 1992), 163–176; Robert D. Bullard, *Dumping in Dixie: Race, Class, and Environmental Quality* (Boulder, CO: Westview Press, 1990); Robert D. Bullard, "Waste and Racism: A Stacked Deck?", *Forum for Applied Research and Public Policy* 8 (1993): 29–45; S. M. Capek, "The 'Environmental Justice' Frame: A Conceptual Discussion and an Application," *Social Problems* 40, no. 1 (1993): 5–24; V. Jordan, "Sins of Omission," *Environmental Action* 11 (1980): 26–27; A. Szasz, *EcoPopulism: Toxic Waste and the Movement for Environmental Justice* (Minneapolis: University of Minnesota Press, 1994); United Church of Christ, Commission for Racial Justice, *Toxic Wastes and Race: A National Report on the Racial and Socio-Economic Characteristics of Communities with Hazardous Waste Sites* (New York: United Church of Christ, 1987); James P. Lester, David W. Allen, and Kelly M. Hill, *Environmental Injustice in the United States: Myths and Realities* (Boulder, CO: Westview Press, 2001).

18. Parks and Roberts, "Environmental and Ecological Justice," 330. Parks and Roberts note that outsiders to the scholarly discussion on justice are usually left with an impression of "philosophical pandemonium . . . a cacophony of discordant philosophical voices . . . incommensurability." Another observer suggests that the pursuit of definitional consensus is a "hopeless and pompous task." Nevertheless, they suggest that a social movement does not need a seamless definition of its core conceptual frame; instead it needs one that motivates people

to act, and one that puts pressure on policymakers who are for a number of reasons averse to being tagged as racist.

19. Hans Kelsen, *Principles of International Law* (New York: Rinehart & Company, 1950).

20. United Nations Charter, Preamble, paragraph 3; cited in Yozo Yokota, "International Justice and the Global Environment," *Journal of International Affairs* 52, no 2 (spring 1999): 585.

21. United Nations Charter, article 2, paragraph 3; quoted in Yokota, "International Justice and the Global Environment," 585.

22. Peter Wenz, *Environmental Justice* (Albany: State University of New York Press, 1988).

23. Tom Tyler, *Why People Obey the Law* (New Haven, CT: Yale University Press, 1990), 178.

24. Willaim J. Aceves, "Critical Jurisprudence and International Legal Scholarship: A Study of Equitable Distribution," *Columbia Journal of Transnational Law* 39 (2001): 391.

25. Aceves, "Critical Jurisprudence and International Legal Scholarship," 391.

26. Aceves, "Critical Jurisprudence and International Legal Scholarship," 392.

27. John Rawls, *A Theory of Justice* (Cambridge, MA: Belknap Press of Harvard University, 1971), 3.

28. David Schlosberg, "Reconceiving Environmental Justice: Global Movements and Political Theories," *Environmental Politics* 13, no. 3 (autumn 2004): 517–540.

29. Iris Marion Young, *Justice and the Politics of Difference* (Princeton, NJ: Princeton University Press, 1990), 1.

30. Nicholas Low and Brendan Gleeson, *Justice, Society and Nature: An Exploration of Political Ecology* (London: Routledge, 1998), 133.

31. Low and Gleeson, *Justice, Society and Nature*, 46, 48, 67.

32. Robert Gordon, "Critical Legal Studies Symposium: Critical Legal Histories," *Stanford Law Review* 36, no. 1/2 (January 1984): 57, 93.

33. Alan C. Lamborn and Joseph Lepgold, *World Politics into the 21st Century*, preliminary ed. (Upper Saddle River, NJ: Prentice Hall, 2003), 486.

34. This highlights the significance of collective-action problems in climate change. Clearly, addressing the prospect of global climate change is consistent with the U.S. national interest. Further, concerns about the free-rider problem have the potential to paralyze negotiations.

35. Michael Grubb, C. Vrolijk, and D. Brack, *The Kyoto Protocol: A Guide and Assessment* (London: Royal Institute of International Affairs, 1999), xxxiii.

36. Roberto Mangabeira Unger, *The Critical Legal Studies Movement* (Cambridge, MA: Harvard University Press, 1986), 89.

37. Taylor, *An Ecological Approach to International Law*, 2.

II

Climate Change, Nature, and Society

5

Climate Change and Arctic Cases: A Normative Exploration of Social-Ecological System Analysis

Amy Lauren Lovecraft

This chapter stands on a precipice overlooking a vast complex system of human and nonhuman animals, vegetation and soils, migration routes, and coastal dynamics that make up the Arctic. It asks the reader to consider, from this vantage and point in time, the future of this social-ecological system at the top of the world. The prospects of this system and its people in light of directional climate change are contingent on choices made by humans that are and will be expressed through practices of social organization, modes of production, and capacities to adapt in concert with the responses of ecological systems. More narrowly, this chapter approaches the contingent futures of the Arctic from the case of Alaska and from an interdisciplinary perspective that has gained traction to analyze complex dynamics of social-ecological systems (SESs). What value does this approach have to political theorists?

Current analyses of SESs have yet to fully theorize the political attributes of the dynamic feedbacks between ecosystems and social systems. This is particularly so in ecological systems experiencing rapid climate change. Human social behaviors are altering many of the factors that determine the fundamental properties of ecological systems, which, in turn, alters the capacity of the environment to sustain human societies as we have come to know them. In the last fifty years humans have changed ecosystems more quickly and extensively than at any comparable period of human history and yet we are just beginning to understand the implications.[1] However, decades of scientific research have established that high latitudes are particularly vulnerable: "The Arctic is now experiencing some of the most rapid and severe climate changes on earth. Over the next 100 years, climate change is expected to accelerate, contributing to major physical, ecological, social, and economic changes, many of which have already begun."[2] The kinds of changes noted by the Arctic Climate Impact Assessment span the totality of human existence.

In other words, the biophysical effects of a changing climate will inherently influence the social dynamics of people living in the Arctic and other locations[3]—it will directly and indirectly affect their lifeworlds. Habermas's *lifeworld* denotes the shared common understandings that individuals and social groups develop over time in their cultures, societies, and personalities. It includes their values, beliefs, and deepest core comprehensions of what it means to be that person, group, or people. All people participate in a lifeworld, mutual participation is at the heart of the matter, and it shapes and socializes us, without our conscious reflection, because it is regularly reaffirmed through the processes we accept as inherent in it. The electoral cycle is part of the American lifeworld. The hunting of bowhead whales is a necessary part of being an Inupiaq or Siberian Yupik person. For most peoples in Arctic nations the phenomenon of climate change is now a piece of their lifeworlds because it has entered the realm of communicative action, the social space where people engage ideas and have meaningful debate. The majority of citizens in the Arctic countries are aware of climate change—its effects have become subject to numerous public agency actions, political pressures, citizen conversations, and governance regimes.

Climate change poses a particular problem for Arctic peoples who livelihoods depend directly (through resource harvest and extraction) or indirectly (through secondary job creation and infrastructure) on natural resources because shifts in the climatic variables that drive ecosystem function such as temperature, ocean currents, and snowfall will alter the ability of an ecosystem to continue to provide the ecosystem services (e.g., timber, clean water, animals, spiritual sites) to the people who depend on them. Major industries operating in the North will also have to adapt to a complex socioeconomic shift as sea ice becomes unpredictable and the inland ice roads unstable as permafrost melts.[4]

As climate change forces people to consider their social (e.g., economic, cultural, political) roles in relation to the environment there is a need to examine the political linkages between social systems and their attendant ecological systems. How will the key drivers of change affect the complex interactions between the two? The preponderance of evidence indicates that rapid changes in the next fifty years in the Arctic will occur. It behooves students of politics to consider what sort of futures will be available to the Arctic and among those which are feasible, equitable, or desired by the majority of citizens. Currently, natural and social scientists propose the concepts of resilience and robustness as desired ends for SESs, but these concepts have yet to be thoroughly articulated in terms of social justice, power relationships, or meanings for the lifeworlds of the humans affected. This chapter examines the effects of rapid climate change on Arctic SESs by focusing on two examples of ecological processes fundamental to human life in the Arctic as we know it—the wildfire disturbance regime of the boreal forest and seasonal sea ice along the northern coasts. These cases will be used to illustrate the analytic power of an SES approach for political scientists but also the fragility of the current resilience and robustness metaphors as goals for social planning, governments, and citizens.

The transformations of these ecological processes will pose hard questions for planners and agencies. How can political theory help us understand the transformational effects of climate change on the Arctic and its consequences? This chapter proposes that in order to fully understand the implications of this process the inquiry must draw from both natural and social sciences in order to develop a framework that serves as a common basis of discussion. I argue that correctly contextualizing and analyzing the effects of climate change on the sociopolitical realm requires a basic knowledge of the structures and functions of ecological subsystems connected to it. Both ecological and social systems respond to a spectrum of controls that operate across a range of temporal and spatial scales. Interdisciplinary efforts to integrate such scales are underway, but no "gold standard" of analysis has been set.[5] This chapter develops by linking two sets of theories representative of promising approaches in both social and natural sciences to grapple with complex human-environment system changes: environmentality and ecosystem services. A framework of analysis is built up from their core premises and the analytic advantages of each are discussed, then the cases are introduced and examined accordingly.

In short, the goal of this chapter is to take steps toward a theoretical framework to accurately examine what the transformation of SESs of the Arctic due to climate change *means* for the people living in them. While the cases are from the Arctic, the relevance of the analysis is broad. For example, island nations, alpine regions, and other locations particularly vulnerable to climate change will experience their own set of problems and opportunities in the coming decades. In terms of political theory, such cases show how the intersecting relationships among politics, institutions, and identities will shift as a result of climate change.

Environmental Political Theory

This chapter works from a broader definition of politics than the classic Laswell definition: "Who gets what, when and how."[6] Such a focus when analyzing natural resource issues can explain distribution patterns, their causes, and implications, but it can prevent us from considering the meaning of institutions, the competition over authority to make decisions tied to human-ecosystem interactions, and the subtle power relationships embodied in the assumptions and objectives of environmental management. As a starting point, this chapter defines politics as "disputes over claims to authority to decide what is, what's right, and what works"[7] in order to expand the examination of cases of SESs affected by climate change beyond describing, for example, what fire gets puts out where and by what method. This chapter proposes that there is a need to understand the *meanings* of the debates and actions tied to climate change in the Arctic. Consequently, both human and nonhuman aspects of the system and the linkages between them that shape meanings, and therefore politics, must be examined. Thus, students of politics bear responsibility to help create a language, a framework, to facilitate communication between natural and social scientists to evaluate the SES approach and its analytical capacity to critique political relationships transforming in light of climate change. In other words, how could such an approach help us understand what the social and ecological transformations will mean to the futures of peoples living in rapidly changing systems? What futures might these transformations present and how might we evaluate them?

This analysis draws on the approach recently refined by Arun Agrawal, *environmentality*, which analyzes how humans' subjective understandings of themselves in relation to their environments change over time due to different technologies of environmental governance. It is helpful to quote his perspective on environmentality at length. He uses the term to denote

a framework of understanding in which technologies of self and power are involved in the creation of new subjects concerned about the environment. There is always a gap between efforts by subjects to fashion themselves anew and the technologies of power that institutional designs seek to consolidate. The realization of particular environmental subjectivities that takes place within this gap is as contingent as it is political. Indeed it is the recognition of contingency that makes it possible to introduce the register of the political in thinking about the creation of the subject.[8]

As he tracks changes in the beliefs and practices of villagers in relation to their forest environment in Kumaon, India, over nearly a century of colonial and postcolonial management, he argues that "new environmental subject positions emerge as a result of involvement in struggles over resources and in relation to new institutions and changing calculations of self-interest and notions of the self. These three conceptual elements—politics, institutions, and identities—are intimately linked."[9] He goes on to note that the success of modern Indian governmental mechanisms to shape the construction of individual and community identities in relation to, and thus their practices toward, their environments in order to create sustainable futures depends on a redefining of political relations, reconfiguring of institutional arrangements, and transformation of environmental subjectivities.[10]

Agrawal's work details these transformations using concepts of power and subjectivity drawn from Foucauldian reasoning. However, while he argues that national and regional governments have caused a change in the average Kumaoni's understanding of himself or herself as an environmental subject, thus subjecting these people to rules of forest management through their buy-ins as those subjects, the actual "biopower" of control is coming from the people themselves. This is not after the fashion of internalized lessons inside the panopticon but through a regulatory process in which subjects participate in rulemaking, environmental monitoring, and rule enforcement. In other words, the self-regulation of forest usage takes place in new "regulatory spaces within localities where social interactions around the environment" occur.[11] Agrawal's larger argument is that the shift to decentralized localized governmental regulatory communities means that while technologies of control can now more easily penetrate into the daily lives, thoughts, and actions of those using the forest resources, this in reality promotes sustainable practices because these communities can more effectively and humanely perform the "control activities" that preserve their environment than centralized state control. He writes of this local autonomy that "the shift from the dynamic of coercion and resistance toward one of involvement in regulatory practices and transformations in environmental subjectivities may be an uncertain process, but it is the goal toward which [these] regulatory communities strive."[12]

Agrawal's exhaustive study demonstrates how the technology of government has promoted sustainable natural resource practices among state subjects with minimal coercive control by the state itself. But this raises timely questions in environmental arenas that are currently caught up in the debate over the use of ecosystems, that are experiencing political resistance to government actions, and that have more recent colonial legacies. In Arctic systems experiencing rapid ecological change and social dislocation, to what end should governments direct their technologies? Whose definition of "sustainability" should be given preference in political struggles, institutional design, and ultimately citizen identity? An analysis of the entire historical context of the Arctic and categorization of its regulatory communities is beyond the scope of this chapter. Here the use of Agrawal's approach is limited to explaining the potential leverage a social-ecological framework can offer political theorists because it can open spaces of power relations for investigation that have previously been hidden and subject them to one of Agrawal's stated aims—to understand what the appropriate goals of environmental management may be.[13] As such, this study uses his concepts to explain the contingent political spaces between social and ecological systems and how the formation of subjects in these spaces depends, in part, on the goals of governance in relation to both systems.

Social-Ecological Systems

Ecological and social systems are so tightly linked that they are often described as "coupled."[14] Interdisciplinary studies researching the characteristics of SESs have begun to develop. Branches of this scholarship strive to understand the feedback between society and ecosystems through analysis of institutional arrangements and management practices,[15] time scales,[16] and how these complex systems may change.[17] Anderies, Janssen, and Ostrom define a SES as "an ecological system intricately linked with and affected by one or more social systems."[18] Both the social and ecological components of a system will have self-organizing independent relationships contained within them (e.g., election cycles are independent of permafrost thaw), as well as having some interactive subsystems (e.g., management limits on harvesting moose depend on how many moose survive to adulthood). In the latter, the interactions are "the subset of social systems in which some of the interdependent relationships among humans are mediated through interactions with biophysical and non-human biological agents." Consider the arrows in figure 5.1. This graphic depicts a SES and its flows. The social system will produce outputs that directly and indirectly affect the ecological system to which it is tied. These outputs can range from collective choices made by units of governance to modes of economic production or the daily decisions of individuals. Furthermore, these choices may be conservation or harvest oriented, they may involve efforts to reduce natural hazards, or they may have no direct relationship to the environment and may be spillover effects of other social processes.[19] The end result, however, remains an impact on the functioning of an ecosystem. The oval encompassing rectangles A and B represents an interactive subsystem, and the narrow rectangles indicate which parts of the flow from social to ecological systems and back are included in this subsystem. Before considering a more detailed discussion of the contingent spaces represented by the rectangles one must address the other direction of the flow, from ecosystem to social system, to more clearly see how social worlds and ecological processes are intertwined.[20]

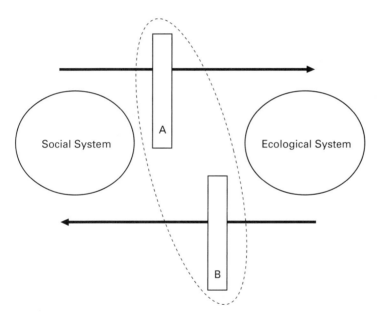

Figure 5.1
The flow of social and ecological outputs from social and ecological systems. **A** represents the slice of diverse social outputs, which include human management of the ecosystem. **B** represents the corresponding slice of ecological services humans manage. The oval surrounding both indicates an interactive subsystem within the social-ecological system (examples are freshwater and the human institutions to procure and manage it, and fire on the landscape and its effects).

Ecosystem Services

A noteworthy analytical tool has been developed in the last few decades that can advance cross-disciplinary comprehension of how societies, individuals, and governments recognize, use, and regulate their environments. The concept of ecosystem services focuses on processes that link ecological and social systems through their effects on human actors.[21] Most simply, these linkages are the interactive flows of outputs from society that affect ecological processes and the resulting benefits and hazards that humans receive from their ecosystems in return. Addressing the latter first, the Millennium Ecosystem Assessment is a synthesis project documenting current global conditions and explaining the relationships between ecosystems and human well-being. Because this chapter seeks to create a framework to analyze power relationships within a changing SES, it is important to review the attributes of an ecological system and how these are perceived by their societies.

Ecosystem services are generally classified as of four types.[22] First, *provisioning* services are the tangible goods people obtain from their natural environment such as food, fiber, genetic resources, fuel, and drinking water Second, ecosystems provide societies with *cultural* services through spiritual opportunities, knowledge systems, social relationships, aesthetic beauty, and other nonmaterial benefits. These first two functions of ecosystems are what most people commonly think of when they perceive what "nature" or "the environment" has to do with people. Third, *regulating* services include the benefits derived from the biochemical processes by the ecosystem itself such as climate regulation from forests, pollination of flowering plants through animal migrations, and freshwater availability affected by glacial melting, soil erosion, and coastal filtration. Finally, the ecological foundations of societies rely on *supporting* services necessary for the production of all the services noted above, such as soil formation, photosynthesis, nutrient cycling, and other fundamental processes that are key to life in ecosystems (see figure 5.2).

Human well-being has five components that are derived from these services, including security, health, social relations, and tangible materials with which to make a good life. These four support the fifth—freedom of choice and action, from which stems the "opportunity to be able to achieve what an individual values doing and being."[23] It is this fifth component that requires careful examination, because when one's freedom to choose and act according to one's values depends on a particular set of ecosystem traits, then that ecosystem is fundamental to well-being. Thus, what are the appropriate institutional goals and

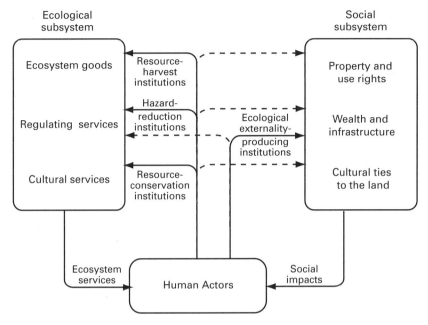

Figure 5.2
Human activities are drivers in both social and ecological systems. This figure expands on figure 5.1 by indicating how societies' institutions are interrelated with the major categories of ecosystem services as well as features in society that are generally tied to human perceptions and management of ecosystems. Normative values and decision making through politics flows upward from human actors to affect the rule sets in each system. (Chapin et al. 2006).

mechanisms to ensure human well-being in relation to Arctic resources? In Alaska, as in most other Arctic states and nations, there is a mixture of highly localized resource-dependent communities along with, though often not spatially near, other communities whose food security, jobs, and social lives are not as tightly tied to the region's ecology.

The Environmental Subject in Interactive Systems

It is crucial to understand these ecosystem services because they are inherently tied to the livelihoods of people and thus their lifeworlds and identities as subjects in relation to both their environments and governments. One way to cut through the interdisciplinary jargon is to look at the interplay between what people perceive themselves to be, how this may change in relation to the environment, how it affects a livelihood, and whether a person/community views this as positive, ultimately creating a particular kind of environmental subject. At this juncture it is important to note that Agrawal uses the phrase "environmental subject" to denote a particular kind of relationship between a person's own subjective understanding of self and the environment. He writes that these people "care about the environment," adding that

For these people the environment is a conceptual category that organizes some of their thinking and a domain in conscious relation to which they perform some of their actions. . . . Further, in considering an actor as an environmental subject I do not demand a purist's version of the environment as necessarily separate from and independent of concerns about material interests, livelihoods, and everyday practices of use and consumption. A desire to protect commonly owned/managed trees and forests, even with the recognition that such protection could enhance one's material self-interest, can be part of an environmental subjectivity. In such situations, self-interest comes to be cognized and realized in terms of the environment.[24]

This definition does beg the question of stewardship, however. Could we argue that people in favor of oil development in Alaska are not environmental subjects? If they self-consciously recognize the environment as the source of their livelihood and seek to protect it just enough to continue further resource extraction does this somehow put them into another category? For purposes of exploring how people situate their identity between social and ecological systems, can recognition of the environment as a "conceptual category" in which they perform actions suffice? Or must we differentiate people who favor the sustainability of ecological systems from those who would exploit them, whether or not they believe they are contributing to sustainability? In this regard, while the development of "environmental subjects" is clearly a valid concept with its own internal logic, it poses a problem for SES analysis because only individuals know their true subjective relationship to the environment and as scholars we can only analyze behaviors. Were we to argue that only behaviors that minimize human impact on the earth are indicative of an environmental subject, we would leave out classes of persons who conceptualize their relationship to their ecosystems as one of stewardship but may not be able to practice sustainable behaviors. Consequently, the institutional dimension of this subjectivity is fundamental to the creation of such subjects and they do not exist, according to Agrawal, without a relationship to sustainable practices that an institution can promote or demand. The existence of environmental subjects, their nature, and resultant behaviors is the crux of this chapter because this subjectivity is contingent on the individual's understanding of his or her SES. Is a person proclaiming "my car does not need a forest" less of an environmental subject than a member of Earth First! because she perceives her relationship to the environment differently, nonetheless still perceiving it? Furthermore, what of indigenous Arctic groups and other examples of place-based peoples whose value systems and livelihoods depend on an Arctic of a particular kind but who may not align with groups traditionally considered "environmental" such as wildlife activists? Wenzel[25] effectively describes this discrepancy in the case of Canada's Inuit, noting that one of the largest challenges to modern Inuit culture is to "alter southern perceptions of the Arctic, including the place of the Inuit in northern ecosystems."[26] The new view often espoused by activists fails to recognize technologies (e.g., snowmobiles and outboard engines) as providing the Inuit an adaptive advantage in light of unpredictable ecosystems (e.g., differences in snowfall and coastal ice formation) to pursue their sociocultural goals of hunting and teaching the young. Should we deny that the Inuit are environmental subjects because of their use of Arctic resources?

This puzzle in determining the true nature of environmental subjects does not, however, prevent the concept from being a constructive functional category to expose political space for analysis between ecosystems and social systems. In particular, it demonstrates the importance of understanding the goals for each system. The interactive subsystem depicted in figure 5.1 encompasses both rectangle A, a slice of the social outputs, and rectangle B, a slice of ecosystem services. These slices correspond to any set of institutions (e.g., rules, regulations, or practices) and beliefs the social system maintains in relation to the ecological system, as well as to the ecological products or results from such actions or beliefs regarding the ecosystem. Subject formation will occur in this oval, due to pressures both within and outside it. Agrawal's definition has come under debate, but its utility in bridging how people control natural resources through rules and how those rules in turn shape the way people conceptualize themselves in relation to the environment is clear.

Resilience, Vulnerability, and Adaptation

The words used to describe human goals for ecological systems have changed over time. Currently, most studies of SESs promote resilience or robustness as their main objectives, which if met, will provide humans with the ecosystem services they have come to expect; they are lately what the system in figure 5.1 should be managed for.

Theories of ecosystem resilience are frequently used to explain the dynamic interactions between people and the ecosystems they inhabit.

Briefly, resilience is the maintenance of fundamental properties of a SES in the face of perturbation:

In ecological systems, resilience lies in the requisite variety of functional groups and the accumulated capital that provides sources for recovery. Resilience within a system is generated by destroying and renewing systems at smaller, faster scales. Ecological resilience is reestablished by the processes that contribute to system "memory"—those involved in regeneration and renewal that connect that system's present to its past and to its neighbors. Management regimes or human activities that remove variability in either space or time will decrease system resilience.[27]

Twinned with resilience theories are those of vulnerability. Vulnerability has to do with the likelihood that the system will experience harm due to exposure to a perturbation or stress/stressor.[28] In times of rapid change, some attributes are more resilient and less vulnerable than others, so different aspects of an ecosystem will respond to stress differently.

How then might we define social resilience? Adger's detailed discussion of the relationships between social and ecological resilience proposes that we cannot deny the "synergistic and coevolutionary relationships" between the two.[29] However, the degree to which any community is dependent on a single ecosystem or single natural resource will make it more vulnerable to impacts on the ecosystem's function or the resource's exploitable abundance. In light of this, he defines *social resilience* as a loose antonym of *vulnerability*: "the ability of communities to withstand external shocks to their social infrastructure." Social resilience is encouraged through institutions, broadly understood as "habitualized behavior and rules and norms that govern society, as well as the more usual notion of formal institutions," which themselves must exhibit resilience.[30] Adger notes that institutional resilience can promote cultural adaptation and the stability of communities but that such resilience is dependent on "cultural context" and "the differing conceptions of human-environment interactions within different knowledge systems."[31] Therefore, the institutions that arise to address interactive SES subsystems will depend on the environmental subjectivity of those creating them and shape these same subjects.

Resilience is a concept popularized in the natural sciences, but recently a different notion of SES management has proposed robustness as a goal. Anderies, Janssen, and Ostrom suggest that we ask what might make systems robust, rather than resilient, and that we should develop robustness by creating a coupled system able to withstand external disturbances and internal stresses without systemwide collapse.[32] Admitting the complications of applying an engineering concept to SESs, the authors create a framework they feel will further analysis of the capacity for robustness in natural resources, their governance systems, and associated infrastructures. Furthermore, they allow for a variety of society-ecology relationships that might be considered robust:

As we explicitly analyze SESs, we distinguish between the collapse or undesirable transformation of a *resource* . . . and the collapse or loss of robustness of the *entire system*. We require that *both* the social and ecological systems collapse before we classify a SES as collapsed and, thus, implicitly define our scale of analysis (or system boundary) to include the human social system and *all* the ecological systems from which it extracts goods and services. . . .

In summary, we suggest that a SES is robust if it prevents the ecological systems upon which it relies from moving into a new domain of attraction that cannot support a human population or that will induce a transition that causes long-term human suffering.[33]

The authors point out several examples of societies where ecosystem collapse of one kind may be desired in order to develop a new kind of ecosystem for social expansion. The shift from mangroves and rice fields to unsustainable shrimp aquaculture in Thailand and Vietnam is considered a choice for robustness because this represents a long-term sustainable "stepping-stone" option of industrial development for the society.[34] Yet, it seems somewhat artificial to describe all socioeconomic systems that maintain some human population and some form of ecological conditions as equally robust or desirable.

Are resilient social systems dependent on resilient ecosystems? Using a SES framework to trace interdependent patterns between any society and any resource(s) is largely a value-neutral enterprise. However, once goals for each system are created, human values direct the formation of the goal as well as the best means to achieve it. These goals are defined and directed through institutions, and these institutions support the social practices that can lead to environmental subjects or in fact resistance to environmental subjecthood.

Human-Environment Interactions: The Role of Institutions

Government technology meets lifeworld in the study of institutions—the systems of rules, norms, and shared strategies, which bind people together to achieve some end.[35] Individuals and groups perceive and respond to their social and ecological environments through a complex web of institutions. Institutional procedures, organizational cultures, and management strategies shape political outcomes because institutions organize and direct social behaviors. During the lifespan of an institution, it creates in its domain certain behavior patterns and epistemologies related to the identification of the rules by people and the ways in which they obey or resist these rules. Peter Brosius, a noted anthropologist studying institutional effects in the environmental movement, elaborates: "Defining themselves as filling particular spaces of discourse and praxis, institutions in effect redefine the space of action; they privilege some forms of action and limit others, they privilege some actors and marginalize others."[36] In other words, concepts like sustainable, community, climate change, and even science have explicit meanings imbued by an institution. The following cases demonstrate how the culture that develops around organized sets of rules affects the formation of politics and identities in relation to governance. In the first case, institutions are coordinated and prevalent; in the second case, there is no sea-ice institutional regime but clusters of rule sets related to ecosystem services. In the former governance, and citizen access to actors and rules that define and regulate wildland fire, is flexible and likely to be able to act quickly to handle changes in the disturbance regime. On the other hand, a lack of overlapping institutional priorities or approaches to sea ice prevents effective mobilization for those affected by its rapid retreat (e.g., polar bears, communities dependent on walrus.)

In short, the potential for institutional development and change in the space between social and ecological systems means potential for politics. The spaces of human-environment interactions demonstrated by rectangles A and B in an interactive subsystem are contingent and subject to institutional colonization, as are the environmental subjectivities that will result from the shaping of these spaces by rules and new social practices in light of climate change. Institutions are shaped by history, embodying "historical trajectories and turning points,"[37] and structure history by offering particular organizational opportunities, perpetuating values related to goals and operational procedures, and cultivating a set of actors within the political system. In the case of the Alaskan indigenous population, there is a history of colonialization and oppression under several governments, and indigenous people in the state only settled claims to land in the 1970s. Consider the interrelated nature of institutions and social-ecological systems depicted in figure 5.2, particularly in light of the two differing case studies that follow.

Interactive Systems in Flux: Wildfire and Sea Ice

One major role of a governing system is to provide its subjects with stability. The root concept for all Western-style democracies is the social contract in which people give up some liberty in order to receive assurances of personal safety. However, this stability typically goes further to provide dependable economic modes of production, an expectation of the ability to participate in regular collective-action decisions such as voting, as well as a general sense of one's culture will be like over time. Climate change threatens to diminish the capacity of governments to provide stability, in the long run, for some human populations. It will continue to increase the unpredictability of weather and earth processes on which people depend for ecosystem services. Considering these two aspects together we must remember that governments also provide values, explicitly and implicitly, in their rules, technologies, practices, and administrative functions.[38]

These two Alaskan cases are demonstrative of some of the biological and geophysical changes occurring in the Arctic that will test the capacity of governments to provide institutions able to manage diverse SESs for the differing ecosystem services desired by various groups. They reveal varying levels of institutional structure tied to the ecosystem as well as the contingent nature of subject formation related to human well-being. The first case addresses the fire regime present in the boreal forests spanning North America, in which institutions have arisen that demonstrate a flexible understanding of the nature of wildland fires in relation to humans. The second case illustrates the fragility of the coastal sea-ice system, which is not subject to any overarching institution and whose services have only recently become a "problem" for humans as the region's charismatic megafauna, such as the polar bear, are threatened with extinction in the wild.

Alaska's Boreal Forests

There are a variety of locations across the globe in which landscapes are wildfire dependent, or fire-adaptive, for ecosystem health[39] and for which extensive periods of fire suppression may lead to more intense, larger, and more frequent fires.[40] Fire on these landscapes serves a variety of important functions, both social and environmental. In biological terms, the biodiversity of the boreal forests of Alaska depend in part on fire. Diverse plant and animal species are one key to resilient ecosystems[41] through the creation of diversity of function. The greater an ecosystem's resilience, the more likely it is to adapt to the stresses placed on it by human populations and their activities. In other words, greater biodiversity means there are more species with traits vital to an ecosystem's function. If some species are lost due to environmental change such as overharvesting or habitat destruction, a richly diverse system might still have enough species with the necessary traits to keep the overall ecosystem functioning.[42] In this manner, humans too are dependent on biodiversity because the natural systems in which we live provide valuable ecosystem services, from foodstuffs and water to textiles and recreation. In boreal forest systems, fire plays a crucial role in providing biodiversity because, among other things, it breaks down organic matter, encourages germination in some plants, and creates a forest mosaic for animals.

Several questions should be asked. What does fire provide for the society inhabiting the boreal-forest SES? How does fire on the landscape influence environmental subjectivity? How might this change with climate unpredictability? Fire impacts are spread over time as well as across different sectors of a human population. Clearly there are negative impacts of fire on a SES such as property destruction, loss of life or livelihood, health effects due to smoke or injury, and loss of some forest resources. However, in Alaskan boreal forests, fire has both short- and long-term benefits. In the short term the positive effects tend to accrue to those who work as firefighters and depend on the fire season for a portion of their wages. In Alaska, where rural residents often have subsistence lifestyles, firefighting wages can account to up to 50 percent of annual cash income.[43] Another short-term effect is the creation of natural fire breaks— once an area burns the vegetation coverage that develops is less likely to burn for the next thirty to sixty years. The longer-term benefits of fire come through the provision of ecosystem services across time. Two to four years after a fire, the vegetation supports mushroom production; two to twenty years after the fire occurs, berries rebound; and ten to thirty years after a fire, moose and furbearers are supported.[44] However, fire can have a more negative effect on caribou, which depend on fire-vulnerable lichens as a winter food source.

It should be noted that these resources are not only important to those practicing subsistence lifestyles for cultural or economic reasons but also to a variety of businesses that cater to food production, hunting and fishing, tourism, and housing development. Over time, the entirety of a boreal forest is regenerated and its products made available to society for these purposes. In short, fires help maintain ecological resilience in fire-adapted forests, and a more resilient system provides a richer array of ecosystem services to humans. Humans, in turn, can gain more for their society's typical goals of survival, capital production, and expansion. But this delicate balance between serving as a functional ecosystem process that results in ecosystem service provision and a natural hazard that can destroy livelihoods is threatened by climate change. The positive and negative attributes of wildfire affect sectors of the human population differently, leading to conflict over how best to manage fire due to jurisdictional questions between state and federal governments, budgetary concerns, access to equipment, and public expectations.

The policy development of wildland firefighting in Alaska from statehood in 1958 until the mid-1980s reflected an aggressive suppression ideology largely consistent with that of the rest of the United States.[45] Three agencies were responsible for coordinating this suppression based on where the fire occurred: the Bureau of Land Management (BLM), the Alaska Department of Natural Resources (DNR), and the U.S. Forest Service (USFS). However, the early 1980s marked a turning point in the state for two reasons. First, the ecological underpinnings of forest health had become more widely understood and the boreal forest was recognized as a fire-adapted system. Second, the costs of total suppression had escalated beyond what the state and federal governments were willing to bear. As a consequence, the government, the fire-management agencies, and the general public began to support a limited policy of managing fires as opposed to completely excluding them from the landscape. The reorganization of priorities and agencies resulted in the 1984 Alaska Consolidated Interagency Fire Management Plan (ACIFMP). Under this new plan the state was divided into thirteen planning areas in which regional planning teams of state and local government officials, land and resource agencies, and regional and village Native representatives[46] would predetermine the level of fire protection needed in their area.

In 1998, the ACIFMP issued a revised plan—now called the Alaska Interagency Wildland Fire Management Plan (AIWFMP)—that consolidated the thirteen plans in order to provide land managers, landowners, and others a single document to reference. In practice fire management in Alaska is now the responsibility of three agencies. The BLM Alaska Fire Service (AFS) manages wildland fire on all Department of the Interior and Native corporation lands and most military land. AFS is further responsible for all lands in the northern part of the state—194 million acres. The State of Alaska Department of Natural Resources Division of Forestry (DNR) manages 150 million acres covering a wide swath in the middle of the state, including the Fairbanks area but not the entirety of the Fairbanks North Star Borough. The U.S. Forest Service does the same in 26 million acres of national forests further south. These agencies coordinate their fire management through the 1998 AIWFMP.

This plan serves as the mechanism for coordinating and operationalizing fire management in Alaska. It clearly defines a relationship between humans and fire:

Fire is now recognized as a critical feature of the natural history of many ecosystems. The evolutionary development of plants and animals has occurred in natural systems where fire was a dominant feature of the environment. Humans occupying an area were also subjected to the natural fire regime, and fire occurrence increased due to human activity. In Alaska, the natural fire regime is characterized by a return interval of 50 to 200 years, depending on the vegetation type, topography, and location.

The goal of this plan is to provide an opportunity, through cooperative planning, for land and resource managers/owners to accomplish fire-related, land-use, and resource management objectives in a cost-efficient manner, consistent with owner, agency, and departmental policies. Management options selected should be ecologically and fiscally sound, operationally feasible, and sufficiently flexible to respond to changes in objectives, fire conditions, land-use patterns, resource information, and technologies.[47]

The plan continues to mix discussion of the importance of fire on the landscape (noting that not all Alaska's regions are fire-adapted, such as the rainforests along the southern coast), with public infrastructure priorities related to fire. In its list of objectives, the first is to "establish wildland fire management option boundaries based upon protection of human life, private property, high-value resources to be protected, and fuel types and their associated fire behavior—not based on administrative boundaries."[48] The second through the sixth objectives relate to fire-suppression responses, review of fire-management needs among land managers and owners, and costs of suppression. The seventh and eighth objectives, again noting the importance of fire on the landscape, are to "minimize adverse environmental impact of fire suppression activities" and "recognize prescribed fire as an important resource management tool to accomplish land and resource management objectives."[49] As a final example of a document with flexible management for a SES, before the body of the document outlines management options and procedures, there are six "General Guidelines." The first states that "the boreal forest and tundra environments are fire-dependent ecosystems, which have evolved in association with fire, and will lose their character, vigor, and faunal and floral diversity if fire is excluded."[50]

This interactive subsystem of fire management policy and fire on the landscape is likely to change due to shifts in climatic variables that increase air and ocean temperatures, as well as creating warmer and drier regional conditions. In interior Alaska, wildland fire appears to be increasing in severity and extent. The fifty-year record shows 60 percent of the largest fires recorded occurring since 1990. Even aspects of boreal systems such as insects and permafrost are tied to fire. Bark beetle outbreaks on the southern margin of the Alaskan boreal forest have increased in part because warming has reduced the length of the beetle's life cycle from two years to one, shifting the balance between the tree and the insect. Insect outbreaks increase the probability of other disturbances natural or human induced (e.g., fire and salvage logging). Permafrost is likely to thaw after fire because combustion and loss of the insulative organic mat makes permafrost temperature more responsive to changes in air temperature. Interior Alaska is estimated to have lost 3 percent of its permafrost area in the last fifty years and is projected to lose all remaining permafrost during the twenty-first century. If this occurs, it will profoundly alter the controls over ecosystem processes (the feedback from regulating to supporting services; see figure 5.2) and will challenge the resilience of the ecological components of the boreal system.[51]

The changes in Alaska's fire regime over time will force new arrangements of institutions and politics as citizens in populated areas demand nearby suppression to reduce smoke, as well as to protect homes, second homes, and businesses in wildland-urban interface locations. Such changes will also create new identities as people learn to change their view of fire on the landscape and the capacity of protective services. Record-breaking fires in Alaska (as well as the adjacent Yukon Territory) in 2004 in both cases have demonstrated positive agency capacity to handle change but also public unwillingness to accept fire nearby.[52] Therefore, this climate change-driven complexity presents governments and environmental subjects a set of "wicked problems" with no optimal solutions.[53] To consider oneself an interactive component of a fire adapted landscape requires physical mobility and a socio-economic culture to match. On the other hand, to develop the landscape without regard to fire is to suppress the hazard that provides for landscape regeneration.

Coastal Sea Ice

Sea ice is another feature of the Arctic that, while it may initially seem as unimportant to human cultural survival as the presence of wildfire, is of critical importance to a variety of SESs in Alaska. If one considers the entire expanse of the Alaskan or Canadian Arctic coastlines, there will be many SESs at varying scales. For example, Barrow and Wales, Alaska, would be different systems but dependent on a similar shared feature— sea ice. In these cases there would be an interactive sea-ice subsystem consisting of those people using that feature of their ecosystem as well as the rules, practices, and identities bound up with this use. However, coastal sea ice itself can be thought of as a system if approached as a pan-Arctic feature of the entire Arctic SES. The interactive subsystem related to sea ice ties together people who rely on this ice for their livelihood and whose choices of livelihood become more dangerous and diminished as the sea ice system changes.

Over the past thirty years the annual average of the extent of sea ice has decreased by approximately one-third, representing an area larger than Norway, Sweden, and Denmark combined.[54] In 2005 and then again in 2007 new record minimums were set.[55] Sea-ice extent in the summer has declined more dramatically with a loss of 15–20 percent of expected cover.[56] Not only has there been less sea ice, but it has also become thinner, with some areas showing a nearly 40 percent reduction in thickness between the 1960s and 1990s[57] as well as an earlier onset of melting. It is not just the loss of sea ice that poses a problem, but the reduced predictability of coastal sea ice as a result of climate change.[58]

These changes produce physical disruptions such as an increase in air temperature, decreased salinity of the ocean's surface layer, and increased coastal erosion, but they also produce social effects, in particular for peoples dependent on sea ice for their economic production and social teaching. Polar bears, several species of seals, and walruses use different aspects of the sea-ice cover as a platform to hunt, rest, and rear their young. These species are all of great importance to indigenous coastal peoples' ways of life. Humans actively use sea ice to travel and to hunt, and as ice thins and retreats in unpredictable ways, people dependent on sea ice both for food provisions and cultural continuity will suffer.[59] It is not only hunting that is endangered from coastal ice dynamics but entire villages. One location dramatically affected is Shishmaref, a northern Alaskan coastal village that has been inhabited for 400 years but is in the process of relocation planning because rising temperatures which reduce sea-ice cover and permit high storm surges to erode the shoreline and undermine the location's homes and infrastructure.[60] There have been repeated efforts by indigenous peoples—largely through the Inuit Circumpolar Council, which represents approximately 150,000 Inuit living in the United States, Canada, Greenland, and Russia—to communicate the urgency of the deteriorating sea-ice conditions and the problems this poses.[61] Federal agencies in the United States and Canada have begun to more widely include Native peoples in their research on coastal sea ice.[62]

A retreat in sea ice will very likely increase the levels of marine transport throughout the Arctic through a longer navigable season and speculation related to mineral extraction, specifically offshore oil and gas.[63] The changes reported in the Arctic summer minimum ice extent represent the most important environmental factor in such a broader context. The prospect of new shipping lanes, extraction of oil and gas resources on previously inaccessible shelves, and problems of national security deemed insignificant in the past due to the inaccessibility of northern regions of many circum-Arctic nations are now resonating within the international arena and have been widely discussed in recent years.[64] The U.S. Minerals Management Service had begun to conduct Outer Continental Shelf lease sales in the Chucki Sea by 2008. This has created extensive political conflict for different groups whose identities and livelihoods are tied to sea ice. For example, the preparation for drilling in such areas involves seismic shoots that penetrate the seafloor and, it is argued, disturb wildlife such as whales and other marine mammals that serve as a vital part of the lives of Arctic coastal peoples. In previous decades, moving ocean pack ice often made drilling uneconomical, but the political, societal, and economic activities afforded by sea ice retreat represent significant opportunities as well as challenges for both Arctic populations and society as a whole. Indicative of the complex nature of this region is the debate over whether or not to list the polar bear as an endangered species. The bears are unlikely to benefit from typical conservation practices, such as further hunting bans (in Alaska hunting is highly restricted with no sport hunting at all), because these rules cannot stop the sea ice retreat that will have the greatest impact on population decline. Diminishing sea ice presents an institutional theorist with a different problem than wildland fire. Without a system of rules currently addressing the lived realities of people on ice, whether Native Alaskan hunters or oil and gas workers on ice islands and roads created to move equipment, one must approach this SES as entirely contingent on the way those involved are now shaping their observations, dialogues, and governance strategies.

Contingency, Agency, and the Environment

Two key elements in a discussion theorizing the interactive spaces between social and ecological systems—populated by the institutions, politics, and identities of human societies—are time and scale. First, disturbances to biological processes that people depend on, such as nutrient cycling in soils, forest cover, water levels, weather patterns, and animal behaviors, may reflect cyclical patterns or be indications of a broader destabilizing trend with science providing answers only over several decades. Second, local observations may remain independent of state, national, or global observations indicating trends not perceived at the local level or failing to indicate trends that local observers have witnessed. The former pushes us to better understand the science of our ecosystems and the human impacts on them. We do know that directional climate change is occurring. The latter pushes us further to examine carefully the multiple disciplinary and nonacademic observations about the effects of climate change on peoples and ecosystems. The chapter proposes a general framework that should help to begin this process.

Will the boreal forests of the North become open grasslands in fifty years? What will the disruption of the fire-disturbance regime mean for the people who live in these forest systems? How can hunters secure food without dependable sea ice? In light of these questions, how should we envision the potential futures available to the Arctic? A window into these questions is offered by Anderies, Janssen, and Ostrom, as well as others who propose robustness or resilience as the desired feature of a SES.[65] Anderies and colleagues propose that SESs should be managed for robustness, noting that they consider a SES robust even if portions of that system's ecology are collapsed beyond the point of no return. So, in practice, societies can transform the natural resources (including entire ecosystems) they use into unsustainable states as long as the society can support its population and there is no "long-term human suffering"—and still be described as robust.[66] Consequently, their work, while providing a skeleton of variables to analyze SESs, does not provide variables that guarantee any particular kind of SES except one that is robust, based on their definition of this concept.

If we apply this to our current task of analyzing the interaction between societies and ecosystems, then, for example, analyzing a fire-dependent socioecological system for robustness may be moot if the resource users and infrastructure providers agree to suppress all fires wherever possible at all times to the detriment of a variety of ecological processes, since the system will still be considered robust because the human population persists in a functioning society. The same could be said of sea-ice coverage, without which it is likely that vast oil reserves could be tapped in the Arctic. On the other hand, if we want to evaluate a SES for the potential of long-term robustness, or even short-term robustness selecting for a valued kind of social-ecological mix, then we have to begin to create variables tied to what people value more closely. What might be done in a SES to make long-term robustness a consideration? This is especially important when a society is unable to afford, or unwilling to pay, the costs (monetary and otherwise) associated with resource depletion and ecosystem collapse that will shift the extraction of goods and services to a different resource base. For example, many rural areas around the world are shifting into unsustainable states as they urbanize. These locations are often in functioning democracies, and the shift is viewed as a positive part of the modernization process. It can also be seen as a strategic step in long-term stability, as noted above in the Vietnamese aquaculture example. But not all societies may choose modernization at the expense of sustainable natural resources or the cultural patterns tied to such resources. In the cases presented above, while many resource users and fire managers have publicly expressed the desire for fewer fires, they have also recognized that they cannot monetarily (or culturally, when one considers what the boreal forest provides people) afford to wipe out this component of their SES. So, how might complex societies manage for the kinds of robustness their inhabitants desire? And what will the inhabitants desire as their subjectivity is shaped by institutions, themselves promoting a kind of robustness?

One element in answering this question is finding a conceptually nonarbitrary way to parse the theory of robustness. Granted, a robust SES must be able to withstand disaster, but what counts as disaster is variable, especially when it comes clothed in an ecologically necessary natural event that provides multiple resources such as wildland fire. If we consider the example of fire-dependent systems, one cannot help but notice that long-term SES health depends on short-term destructive forces. Assuming they do not want a wholesale shift in their environs such as converting the boreal forest to farmland, societies must balance short-range vulnerabilities to fire and its attendant socially inefficient results (e.g., smoke, altering of hiking trails, loss of some tourism, loss of property) with long-term ecosystem-service delivery (e.g., moose, scenic forests, healthy forests for timber, berries). Considering these trade-offs, we can conceptualize both weak and strong societies as well as weak and strong ecosystems. For example, do we move everyone out of the boreal forest in order for fire to run its course? Do we suppress fire to maximize clear air and home development? If robustness is a useful concept because it "emphasizes the cost-benefit trade-offs associated with systems designed to cope with uncertainty,"[67] then these trade-offs

	Max Ecosystem	Min Ecosystem
Max Society	(1) This is the "best" option. We could define the attributes of this system as having high quality of living in society coupled to a flourishing ecosystem with resilient capacity to withstand external shocks (disasters or climate change) and changes in demands on ecosystem services.	(2) Nearly unsustainable ecosystem but flourishing social system. Often locations where human advances in living have come at the expense of resource destruction/ depletion and unsustainable ecosystems. Locations where society is entirely dependent on non-local natural resources such as major urban centers.
Min Society	(3) Precariously sustainable society (one with narrow margins of change) but coupled to a flourishing ecosystem. Perhaps nomadic groups, small bands of hunter gatherers. Very rural sparsely populated locations.	(4) The "worst" option with unsustainable societies in barren landscapes. Examples might be locations struck by natural disasters that wipe out both societies and ecosystems. Locations with extreme heat or cold.

Figure 5.3
Robustness typology for coupled SES systems.

should be to some extent reflected in the application of the concept. A fourfold model of SES robustness might go a long way in helping to practically understand some of the trade-offs societies are willing to make and for what reasons (see figure 5.3).

Around the globe, we can find examples of resilient ecosystems coupled with flourishing societies, as well as the opposite: degraded ecosystems coupled with barely sustainable human populations. However, examples of highly successful societies located in virtually barren ecosystems and rich ecosystems with only a precariously sustainable society also exist. This raises the question of choice in society: Do all societies have the capacity to recognize the trade-offs they face? In other words, one cannot assume that all societies are able to cope with uncertainty. Furthermore, the societies must first be able to understand that their future way of life may *be* uncertain. Are they all equally able to develop the capacity to make long-range decisions related to the desired outcomes from these trade-offs? If the answer to both of these questions is no, can we still call the SES robust? When societies can recognize future uncertainty but are unable to design socially acceptable means of coping with it, either for external or internal reasons, can we measure them against societies with this capacity? These questions leave room for further research into the linkages between social choices and ecological futures.

Conclusion: Problems and Opportunities

Viewing politics as "struggles over claims to authority to decide what is, what is right, and what works" means that politics in Arctic SESs stem in part from socially constructed debates over the nature, meaning, and administration of fire on the landscape or sea-ice cover. These debates have shaped institutions that now have authority to make decisions affecting SESs across geographic and time scales. It is thus not a stretch to argue that the politics surrounding these institutions is partially driven by forces in the ecosystems. Because, in the United States, these politics play out in the social system of a polyarchal democracy, the tensions experienced within the system are currently not precursors to either social collapse or ecosystem ruin for the majority of citizens. But both are possible in this century for smaller numbers of people whose identities are tied to SESs but for whom politics and institutions do not appear to offer any short-term help; nor are their numbers sufficient to effect electoral change. The major challenge for the pan-Arctic SES as a whole is its ability to form and implement the long-range planning capacity to research and make decisions related to climate change, population growth, and shifting demands on natural systems. Furthermore, this challenge cannot solely be answered through "politics as usual," which reinforces some environmental subjectivity at the expense of others.

Human well-being depends on the capacity of institutions to develop in such a way as to offer genuine choices to those most affected by climate change, because the definition of human well-being includes freedom of choice and action. A person must have room to develop an environmental subjectivity congruent with his or her opportunities in order to be able to achieve what he or she values doing and being. Likewise opportunities (e.g., access to diverse ecosystems) must be available for the development of a range of subjectivities. But the directional stresses facing the Arctic system present a situation wherein indigenous cultures and people who choose to live subsistence lifestyles will be unable to choose activities or life patterns that are fundamental to cultural survival because the environment they need will be dramatically altered. This space of institutional development and subject creation is currently contingent on choices made not only by those living in the Arctic—but by those far to the south of it.

Notes

I am indebted to Dr. Terry Chapin (University of Alaska Fairbanks) for his exchange of ideas and support; to my research partner Dr. Hajo Eicken (also UAF), for contributing greatly to the discussion on sea ice, and to the Arctic System Science program at the National Science Foundation for their funding of the Human-Fire Interaction Project at the University of Alaska (grant OPP-0328282). I also thank the NSF EPSCoR program. Lastly, my thanks to the *Proceedings of the National Academy of Sciences* and *Quebec Studies* publications for their kind permissions.

1. F. S. Chapin III, A. L. Lovecraft, E. S. Zavaleta, J. Nelson, M. D. Robards, G. P. Kofinas, S. F. Trainor, G. D. Peterson, H. P. Huntington, and R. L. Naylor, 2006. "Policy strategies to address sustainability of Alaskan boreal forests in response to a directionally changing climate," *Proceedings of the National Academy of Sciences* 103(2006): 16637–16643; Millennium Ecosystem Assessment (MEA), *Ecosystems and Human Well-Being: Synthesis* (Washington, DC: Island Press, 2005).

2. ACIA, *Impacts of a Warming Arctic: Arctic Climate Impact Assessment* (New York: Cambridge University Press, 2004), 10.

3. L. D. Hinzman, Bettez, N. D., Bolton, W. R., Chapin, F. S., III, Dyurgerov, M. B., Fastie, C. L., Griffith, B., Hollister, R. D., Hope, A., Huntington, H. P., Jensen, A. M., Jia, G. J., Jorgenson, T., Kane, D. L., Klein, D. R., Kofinas, G., Lynch, A. H., Lloyd, A. H., McGuire, A. D., Nelson, F. E., Nolan, M., Oechel, W. C., Osterkamp, T. E., Racine, C. H., Romanovsky, V. E., Stone, R. S., Stow, D. A., Sturm, M., Tweedie, C. E., Vourlitis, G. L., Walker, M. D., Walker, D. A., Webber, P. J., Welker, J. M., Winker, K. S. & Yoshikawa, K., "Evidence and Implications of Recent Climate Change in Northern Alaska and Other Arctic Regions," *Climatic Change* 72 (2005): 251–298.

4. Arctic Council, *Arctic Marine Strategic Plan*, 2004, www.pame.is.

5. D. W. Cash and S. C. Moser, "Linking Local and Global Scales: Designing Dynamic Assessment and Management Processes," *Global Environmental Change* 10 (2000): 109–120; Chapin et al., "Policy Strategies to Address Sustainability of Alaskan Boreal Forests in Response to a Directionally Changing Climate."; W. C. Clark and N. Dickson, "Sustainability Science: The Emerging Research Program," *Proceedings of the National Academy of Sciences* 100, no. 14 (2003): 8059–8061; Neil Adger, "Social and Ecological Resilience: Are They Related?", *Progress in Human Geography* 24, no. 3 (2000): 347–364.

6. Harold Laswell, *Politics: Who Gets What, When, and How* (New York: McGraw-Hill, 1936).

7. D. V. Edwards and A. Lippucci, *Practicing American Politics* (New York: Worth Publishers, 1998).

8. Arun Agrawal, *Environmentality: Technologies of Government and the Making of Subjects* (Durham, NC: Duke University Press, 2005), 166.

9. Agrawal, *Environmentality*, 3.

10. Agrawal, *Environmentality*, 7.

11. Agrawal, *Environmentality*, 6–7.

12. Agrawal, *Environmentality*, 163.

13. Agrawal, *Environmentality*, 8.

14. F. Berkes and C. Folke, "Investing in Cultural Capital for Sustainable Use of Natural Capital," in A. M. Jansson, M. Hammer, C. Folke, and R. Constanza, eds., *Investing in Natural Capital: The Ecological Economics Approach to Sustainability* (Washington, DC: Island Press, 1994).

15. E. Ostrom, *Governing the Commons: The Evolution of Institutions for Collective Action* (New York: Cambridge University Press, 1990); S. Hanna and S. Jentoft, "Human Use of the Natural Environment: An Overview of Social and Economic Dimensions," in Susan Hanna, Carle Folke, and Karl-Goran Maler, eds., *Rights to Nature: Ecological, Economic, Cultural, and Political Principles of Institutions for the Environment* (Washington, DC: Island Press, 1996); F. Berkes and C. Folke, "Investing in Cultural Capital for Sustainable Use of Natural Capital"; J. M. Anderies, M. A. Janssen, and E. Ostrom, "A Framework to Analyze the Robustness of Social-Ecological Systems from an Institutional Perspective," *Ecology and Society* 9, no. 1 (2004): 18.

16. T. Abel, "Complex Adaptive Systems, Evolutionism, and Ecology within Anthropology: Interdisciplinary Research for Understanding Cultural and Ecological Dynamics," *Georgia Journal of Ecological Anthropology* 2 (1998): 6–29; H. T. Odum, *Systems Ecology* (New York: Wiley, 1983); W. E. Odum, E. P. Odum, and H. T. Odum, "Nature's Pulsing Paradigm," *Estuaries* 18, no. 4 (1995): 547–555.

17. L. H. Gunderson and C. S. Holling, eds., *Panarchy: Understanding Transformations in Human and Natural Systems* (Washington, DC: Island Press, 2002); R. L. Constanza, L. Wainger, C. Folke, and K. Maler, "Modeling Complex Ecological Economic Systems: Toward an Evolutionary, Dynamic Understanding of People and Nature," *BioScience* 43, no. 8 (1993): 545–555.

18. Anderies, Janssen, and Ostrom, "A Framework to Analyze the Robustness of Social-Ecological Systems from an Institutional Perspective."

19. F. S. Chapin et al., "Policy Strategies to Address Sustainability of Alaskan Boreal Forests in Response to a Directionally Changing Climate."

20. For an example of this concept applied to transboundary freshwater systems in North America, see A. Lovecraft, "Bridging the Biophysical and Social in Transboundary Water Governance: Quebec and its Neighbors," *Quebec Studies* 42 (2007), 133–140.

21. Oran R. Young, *The Institutional Dimensions of Environmental Change: Fit, Interplay, and Scale* (Cambridge, MA: MIT Press, 2002); T. Dietz, E. Ostrom, and P. C. Stern, "The Struggle to Govern the Commons," *Science* 302 (2003): 1907–1912.

22. Millennium Ecosystem Assessment, *Ecosystems and Human Well-Being*.

23. Millennium Ecosystem Assessment, *Ecosystems and Human Well-Being*, vi.

24. Arun Agrawal, "Environmentality: Community, Intimate Government, and the Making of Environmental Subjects in Kumaon, India," *Current Anthropology* 46, no. 2 (2005): 161–190 (quote on 162).

25. George W. Wenzel, "Warming the Arctic: Environmentalism and Canadian Arctic," in D. Peterson and D. Johnson, eds., *Human Ecology and Climate Change* (Bristol, PA: Taylor & Francis, 1995), 169–184.

26. Wenzel, "Warming the Arctic," 175.

27. Gunderson and Holling, *Panarchy.*

28. B. L. Turner II, R. E. Kasperson, P. A. Matson, J. J. McCarthy, R. W. Corell, L. Christensen, N. Eckley, J. X. Kasperson, A. Luers, M. L. Martello, C. Polsky, A. Pulsipher, and A. Schiller, *Proceedings of the National Academy of Sciences, USA* 100 (2003): 8074–8079.

29. Neil Adger, "Social and Ecological Resilience: Are They Related?", *Progress in Human Geography* 24, no. 3 (2000): 347–364 (see especially 350).

30. Adger, "Social and Ecological Resilience: Are They Related?", 348.

31. Adger, "Social and Ecological Resilience: Are They Related?", 351.

32. Anderies, Janssen, and Ostrom, "A Framework to Analyze the Robustness of Social-Ecological Systems from an Institutional Perspective."

33. Anderies, Janssen, and Ostrom, "A Framework to Analyze the Robustness of Social-Ecological Systems from an Institutional Perspective," 7; italics in original.

34. Anderies, Janssen, and Ostrom, "A Framework to Analyze the Robustness of Social-Ecological Systems from an Institutional Perspective"; L. Lebel, N. H. Tri, A. Saengnoree, S. Pasong, U. Buatama, and L. K. Thoa, "Industrial Transformation and Shrimp Aquaculture in Thailand and Vietnam: Pathways to Ecological, Social, and Economic Sustainability?", *Ambio* 31, no. 4 (2002): 311–322.

35. Oran R. Young, *The Institutional Dimensions of Environmental Change.*

36. J. Peter Brosius, "Green Dots, Pink Hearts: Displacing Politics from the Malaysian Rain Forest," *American Anthropologist* 101 (1999): 36–57 (quote on 50).

37. Robert Putnam, *Making Democracy Work* (Princeton, NJ: Princeton University Press, 1993), 8.

38. John Kirlin, "What Government Must Do Well: Creating Value for Society," *Journal of Public Administration Research and Theory* 6, no. 1 (1996): 161–185.

39. D. A. Wardell, T. T. Nielsen, K. Rasmussen, and C. Mbow, "Fire History, Fire Regimes and Fire Management in West Africa: An Overview," in J. G. Goldammer and C. de Ronde, eds., *Wildland Fire Management Handbook for Sub-Sahara Africa* (Freiburg: Oneworldbooks, 2004); S. C. Chapin, S. T. Rupp, A. M. Starfield, L. DeWilde, E. Zavaleta, N. Fresco, J. Henkelman, and D. A. McGuire, "Planning for Resilience: Modeling Change in Human-Fire Interactions in the Alaskan Boreal Forest," *Frontiers in Ecology* 1, no. 5 (2003): 255–

261; A. N. Andersen, G. D. Cook, and R. J. Williams, eds., *Fire in Tropical Savannas: The Kapalga Experiment* (New York: Springer, 2003); Seth Reice, *The Silver Lining: The Benefits of Natural Disasters* (Princeton, NJ: Princeton University Press, 2001); R. J. Whelan, *The Ecology of Fire* (New York: Cambridge University Press, 1995).

40. G. J. Busenberg, "Adaptive Policy Design for the Management of Wildfire Hazards," *American Behavioral Scientist* 48, no. 3 (2004): 314–326; S. J. Pyne, *Fire in America: A Cultural History of Wildland and Rural Fire* (Princeton, NJ: Princeton University Press, 1982); D. Carle, *Burning Questions: America's fight with nature's fire* (Westport, CT: Praeger, 2002); C. Dennis, "Burning Issues," *Nature* 421 (January 2003): 204–206; R. N. Sampson, "Primed for a Firestorm," *Forum for Applied Research and Public Policy* 14, no. 1 (1999): 20–25.

41. Constanza et al., "Modeling Complex Ecological Economic Systems."

42. S. C. Chapin, B. H. Walker, R. J. Hobbs, D. U. Hooper, J. H. Lawton, O. E. Sala, and D. H. Tilman, "Biotic Control over the Functioning of Ecosystems," *Science* 277 (1997): 500–504; Seth Reice, *The Silver Lining*.

43. Chapin et al., "Planning for Resilience."

44. Chapin et al., "Planning for Resilience."

45. D. C. Natcher, "Implications of Fire Policy on Native Land Use in the Yukon Flats, Alaska," *Human Ecology* 32, no. 4 (2004): 421–441.

46. Briefly, when Alaska achieved statehood the BLM ceded over 100 million acres of federally owned land to the new state, the U.S. Park Service, and the U.S. Fish and Wildlife Service. Later the 1971 Alaska Native Claims Settlement Act (ANCSA) was enacted by Congress to settle aboriginal land claims by Natives and Native groups of Alaska, creating a system of land ownership for thirteen local and regional Native corporations that could realize economic benefits. Under ANCSA the BLM ceded a further forty-four million acres to the Native corporations. Consequently, these Native authorities have contractual relationships with the Alaska Fire Service for wildland fire management.

47. U.S. Department of the Interior (USDI), *Alaska Consolidated Interagency Fire Management Plan*, operational draft (Fairbanks, AK: U.S. Bureau of Land Management, 1998), 12.

48. USDI, *Alaska Consolidated Interagency Fire Management Plan*, 12–13.

49. USDI, *Alaska Consolidated Interagency Fire Management Plan*, 13.

50. USDI, *Alaska Consolidated Interagency Fire Management Plan*, 14.

51. Chapin et al., "Policy Strategies to Address Sustainability of Alaskan Boreal Forests in Response to a Directionally Changing Climate."

52. A. L. Lovecraft and S. F. Trainor, "Organizational Learning and Policy Change in Wildland Fire Agencies: Cases of Uncharacteristic Wildfires in Alaska and Yukon Territory," unpublished manuscript, 2006.

53. F. S. Chapin, , S. F. Trainor, O. Huntington, A. L. Lovecraft, E. Zavaleta, D. C. Natcher, A. D. McGuire, J. L. Nelson, L. Ray, M. Calef, N. Fresco, H. Huntington, S. Rupp, L. DeWilde, R. L. Nayor, "Increasing Wildfire in Alaska's

Boreal Forest: Causes, Consequences, and Pathways to Solutions of a Wicked Problem" *BioScience* (2008) (in press).

54. ACIA, *Impacts of a Warming Arctic: Arctic Climate Impact Assessment*; J. C. Comiso, "A Rapidly Declining Perennial Sea Ice Cover in the Arctic," *Geophysical Research Letters* 29 (2002): 1956.

55. J. C. Stroeve, M. C. Serreze, F. Fetterer, T. Arbetter, W. Meier, J. Maslanik, and K. Knowles, "Tracking the Arctic's Shrinking Ice Cover: Another Extreme September Minimum in 2004," *Geophysical Research Letters* 32 (2005): L04501; Serreze, M. C., M. M. Holland, and J. Stroeve, "Perspectives on the Arctic's shrinking sea-ice cover," *Science 315* (2007): 1533–1536.

56. ACIA, *Impacts of a Warming Arctic.*

57. ACIA, *Impacts of a Warming Arctic.*

58. I. Krupnik and D. Jolly, *The Earth is Faster Now: Indigenous Observations of Arctic Environmental Change* (Fairbanks, AK: Arctic Research Consortium of the United States, 2002).

59. S. Fox, *When the Weather Is Uggianaqtuq: Inuit Observations of Environmental Change*, CD-ROM (Boulder: University of Colorado Geography Department Cartography Lab, 2003).

60. NOAA, "Arctic Change: A Near-Realtime Arctic Change Indicator," 2006 (updated November), http://www.arctic.noaa.gov/detect/human-shishmaref.shtml.

61. Sheila Watt-Cloutier, "Inuit Circumpolar Conference Testimony," U.S. Senate Committee on Commerce, Science, and Transportation, Washington, DC, September 15, 2004, http://www.ciel.org/Publications/McCainHearingSpeech15Sept04.pdf.

62. NOAA, "Changes in Arctic Sea Ice over the Past 50 Years: Bridging the Knowledge Gap between Scientific Community and Alaska Native Community," Executive Summary from the Marine Mammal Commission Workshop on the Impacts of Changes in Sea Ice and Other Environmental Parameters in the Arctic, 2000, http://www.arctic.noaa.gov/workshop_summary.html.

63. L. W. Brigham, "Thinking About the Arctic's Future: Scenarios for 2040," *The Futurist* (Sept/Oct 2007): 27–34.

64. Arctic Council, *Arctic Marine Strategic Plan.*

65. At this time, only Anderies, Janssen, and Ostrom have proposed a framework for a social-ecological system goal. Studies in resilience still tend to discuss either social or ecological resilience, or compare them (see Adger, "Social and Ecological Resilience: Are They Related?"), but have not addressed the goal of the coupled SES itself. Consequently, I focus on Anderies, Janssen, and Ostrom's model.

66. Anderies, Janssen, and Ostrom, "A Framework to Analyze the Robustness of Social-Ecological Systems from an Institutional Perspective."

67. Anderies, Janssen, and Ostrom, "A Framework to Analyze the Robustness of Social-Ecological Systems from an Institutional Perspective," 1.

6

Climatologies as Social Critique: The Social Construction/Creation of Global Warming, Global Dimming, and Global Cooling

Timothy W. Luke

This study asks why the inchoate workings of contemporary industrial production and consumption leave behind huge noxious by-products, like carbon dioxide, methane, nitrous oxide, and chlorofluorocarbons, to create "global cooling," "global dimming," or "global warming." It also asks how these trends slowly are reconstructing nature in enduring ways, which now are openly addressed by the sciences of climatology as a form of social criticism. Some climatologists accept this social engagement; yet, their analyses also imply these changes are so rapid, profound, and fundamental that a new kind of environment, which some are identifying as "socionature," "technonature," or "urbanatura," is arising from their interactions. And, despite constant surveillance by scientific experts, these climate changes seem to confound most human responses due to their remarkable unpredictability and vast scope. To survey these changes, then, this critique will reconsider the social creation, as well as the social construction, of global warming, dimming, and cooling.

No environment exists independently of the organisms it envelops, and the human life forms that the earth's environments encircle are, intentionally and unintentionally, profoundly altering those environments to maintain an unsustainable economy. Consequently, common notions about what the earth's "environment" has been understood as, and how "environmentalists" could organize for its defense, now require foundational changes once one recognizes how much widespread anthropogenic processes actually are warming, dimming, and/or cooling the planet.[1]

Global Warming

Around two million years ago, the protohuman lines of the genus *Homo*, which gave rise to *Homo sapiens*, appear in the fossil record. As they evolved, a remarkable decrease in global temperatures around 900,000 years ago initiated a fairly regular pattern of repeated ice ages that each last around 100,000 years, but alternate with shorter, warmer eras of 8,000 to 40,000 years. Within this cycle, the last ice age ended around 18,000 years ago, except for the Younger-Dryas event that suddenly returned the earth to ice-age-like temperatures for only around 200 years. Just as remarkably, rewarming occurred quickly, with temperatures in Greenland, for example, rising 10°F in less than ten years.[2]

Many climatological studies accept the historical convention that the Industrial Revolution began in the eighteenth century as steam engines and growing cities led to tremendous increases in the consumption of coal, wood, and biomass fuels to generate the energy needed for modern industrial life. Geological, botanical, and oceanographic evidence also reveals increasing levels of carbon dioxide and other greenhouse gases beginning in that period. The reimagination of the earth as "infrastructural systems"[3] and spaces is noted, in turn, as early as 1824 in a scientific study by Jean-Baptiste-Joseph Fourier. His "General Remarks on the Temperature of the Terrestrial Globe and Planetary Spaces" recast the biophysics of atmospheric chemistry, solar radiation, and terrestrial temperature as a structure interoperating like a giant glass dome to generate the warmth needed to sustain the biosphere, with all its human and nonhuman inhabitants.[4]

Fourier's work was conducted during the last decades of the "Little Ice Age" of 1300–1900, and his interest in how global warmth could be maintained is not surprising. Similarly, the Swedish chemist Svante August Arrhenius kept with this positive outlook on global warming in a study that links the earth's cycles of ice ages and warmer interglacial eras to variations in the levels of carbon dioxide in the planet's atmosphere. Like John Tyndall (who coined the term *greenhouse gases*) in the 1860s, Arrhenius recognized that water vapor and ozone also help absorb and retain heat, but he went beyond Tyndall to argue that increasing carbon dioxide emissions could enhance this greenhouse effect, sustain global warming, and improve weather conditions for humanity. A meteorologist working Great Britain, Guy S. Callendar, seconds Arrhenius's work by documenting a 1° rise in the earth's temperature from 1880 to 1934. He tied this increase to human fossil fuel use, and suggests that another 2°F increase in average planetary temperatures would occur by the 2030s. His conclusions about these trends were positive, because Callendar believed such global warming would improve agriculture, postpone the world's periodic reglaciation, and maintain better living conditions.[5]

Implicitly, then, the traditional boundary properties of concepts like nature/society, city/country, and urban/rural were being challenged not long after industrialization began. Likewise, the property boundaries of cities were recognized as being exceeded, and then sublated, by their noxious by-products as well as their beneficent products. As fossil fuel wastes accumulated in the atmosphere, the earth itself was recast by the scientific imagination as artifice, architecture, or artifact, because greenhouse gassing was crystallizing anthropogenic wastes with atmospheric chemistries as essentially a terrestrial greenhouse. The urban and natural quickly were (con)fused, and the re-terraforming hybridities of "urbanatura" arguably arose along with the smoke and ashes.

After a rush of studies, experiments, and commissions stretching back into the 1970s to examine chlorofluorocarbons, ozone, and carbon dioxide, scientists working with the World Meteorological Organization (WMO), the United Nations Environmental Program (UNEP), and the International Council on Scientific Unions (ICSU) met in 1980 to express concern about rising carbon dioxide emissions in particular. After several years of additional study, the United Nations agreed to sponsor a task force focused on climate change to monitor trends like global warming, dimming, and cooling. The Intergovernmental Panel on Climate Change (IPCC), then, was organized by the UN in 1988 as a joint action group of the UNEP and the WMO. As a forum for contending scientific perspectives as well as a creature of the world's key International Governmental Organizations (IGOs), as Maslin asserts, the IPCC is meant to provide "the continued assessment of the state of knowledge on the various aspects of climate change, including scientific, environmental, and socio-economic impacts and response strategies. The IPCC is recognized as the most authoritative scientific and technical voice on climate change, and its assessments have had a profound influence on the negotiators of the United Nations Framework Convention on Climate Change (UNFCC) and its on-going Kyoto Protocol."[6]

Given this institutional position, the IPCC's operations are conducted by a joint task force plus three working groups, all cochaired by one representative each from a developed country and a developing country:

Working Group I assesses the scientific aspects of the climate system and climate change; Working Group II addresses the vulnerability of human and natural systems to climate change, the negative and positive consequences of climate change, and the options for adapting to them; and Working Group III assesses options for limiting greenhouse gas emissions and otherwise mitigating climate change, as well as economic issues. Hence the IPCC also provides governments with scientific, technical, and socio-economic information relevant to evaluating the risks and to developing a response to global climate change.[7]

While the missions of the three working groups obviously overlap, and the cochairing arrangements institutionalize the inequality of nations in the IPCC's workings, these bodies do struggle to integrate the findings of hundreds of scientific experts from dozen of nations to provide an ongoing assessment of greenhouse gases and their role in global warming. Following on the work of Arrhenius and Callendar, a sustained effort to monitor this trend by generating new continuously collected data at the Mauna Loa Observatory in Hawaii was initiated in 1958. Its ongoing annual measurements showed carbon dioxide levels in the earth's atmosphere rising rapidly by 11 percent in only four decades.[8]

Global Cooling

Climate variation usually occurs on a geological rather than a historical time scale. Paleoclimatology, geology, paleontology, and oceanography have discovered and verified evidence of many previous global warming and global cooling periods. Some appear relatively minuscule and cyclical, others seem truly rapid and episodic, and a few are extraordinarily catastrophic and long-lived. Over the past two millennia, a 400-year-long episode of global warming occurred from 900 to 1300 AD, which coincided with the medieval era in Europe and the Sung to Yuan dynasties in China. Likewise, a 600-year-old "Little Ice Age" took place from 1300 to 1900 AD, which was the backdrop for the rise of industrial capitalism in Europe. Aside from the comparatively brief Medieval Warming period, global temperatures actually were cooling from 1000 to 1900 AD by nearly 2.7°F to 3.6°F. Of course, the twentieth century was the warmest century in the last millennium, the 1990s stood out as its warmest decade, and 1998 in the Northern Hemisphere was its warmest year since 1000.[9] Temperatures over the past 10,000 years, since the last Ice Age ended around 12,000 years ago, during our Holocene Era have varied as much as 9°F to 14.4°F in as little time as 1,500 years.[10] And this latest epoch, which has also seen the rise of settled urban civilization, has been the longest warm and relatively stable period over the last 400 millennia. Nevertheless, there have been moments of global cooling as well, especially during the early years of carbon-emitting industrial capitalism. During the 1970s, climatologists actually were predicting a prolonged period of global cooling after nearly forty years of expanding glaciation, cooler temperatures, and harsh winters.[11] Even though other scientists were suggesting that global warming would become more prevalent, the albedo, or reflective, effects of industrial pollution did permit scientists to credibly argue a case for global cooling.

Global cooling in the geological record can be tied to multiple sources, ranging from the extraterrestrial, which have been caused by meteor strikes, to the terrestrial, which seem to be caused by ocean current changes due to nonanthropogenic global warming. The effects of global warming are not uniform, and most of this phenomenon has been found between 40° and 70° North latitude. Yet, some areas over land in these latitudes and in the North Atlantic Ocean actually have cooled in the past few decades.[12]

Here the earth's cryosphere plays an intriguing role. The massive ice concentrations over the Arctic, Greenland, and Antarctica, if they melt, will cause sea levels to rise. Their thinning along with glacial melting in alpine regions already accounted for about two to five centimeters of ocean-level increases in the twentieth century. American and Russian military submarine observations suggest the ice drafts in the Arctic Ocean's deep waters were over a meter thinner in the 1990s than in the 1950s.[13] Moreover, since 2000, larger expanses of open water have been appearing in the Arctic Ocean, even during the winter. This thinning of the cryosphere is significant to the degree that the expanses of ice covering parts of the planet are a major contributor to the planet's albedo effects. A considerable amount of solar radiation is reflected back into space, keeping the earth partially cool. As open water, soil, or vegetation replaces ice in current cryosphere zones, it will absorb more radiation and could advance global warming.

However, anthropogenic atmospheric changes are not only connected to global warming. Industrial greenhouse gassing also is associated with more particulate and aerosol emissions, and their regional concentrations account for local cooling by reflecting considerable amounts of solar radiation back into the atmosphere. Likewise, water vapor caused by warmer temperatures and shifting weather patterns also is reflecting more solar radiation back into space. Hence atmospheric concentrations of industrial aerosols, particulates, or other dense greenhouse gases, according to some models, are quite likely to advance global cooling along with global warming.[14]

Global Dimming

Even though it does not yet have the widespread attention given to global warming or cooling, another phenomenon, which also seems associated with radical atmospheric changes, is global dimming. Studies of global warming and global dimming still are contested in the scientific community, but careful observations done since the 1960s show drastic decreases in the amount of solar radiation falling on the earth's surface. Many scientists continue to dismiss these observations as inaccurate, improbable, or even impossible. Yet, longitudinal studies from the 1950s to the 1990s do show that the amount of solar radiation reaching the surface of the planet has declined on average by .23 to .32 percent per year.[15]

These effects are not only caused by a decrease in solar radiation itself, because the sun's output during this time basically has remained constant. Instead it appears that levels of human-made particulates and chemical compounds are increasing very rapidly. As they rise into the atmosphere, they help form clouds that are thicker, darker, and longer lasting. This cloud cover then reflects more solar radiation back into space. Increases in fossil fuel use, and a widespread application of aerosol chemicals, combine with condensed water and act as dimming pollutants. Likewise, jet airplane traffic leaves contrails in the atmosphere that contribute to these effects. In fact, this one single factor did much to trigger more systematic studies of global dimming. For three days after the September 2001 terrorist attacks in the United States, almost all jet aircraft going in and out of North America were grounded. Observations around the Northern Hemisphere revealed an immediate temperature increase of over 1°C. Such changes usually occur over many years, but this temperature change took place in just over seventy-two hours.[16]

Measurements in the field had documented this trend since the mid-1980s in several places, but they were ignored during the more intense debates over global warming and the threat of a nuclear winter. Atsumu Ohmura at the Swiss Federal Institute of Technology first identified global dimming in 1985 when he found that the level of solar radiation reaching the earth's surface seemed to have fallen by over 10 percent in three decades. He published his findings in 1989, but they were largely ignored.[17] Indeed, the Intergovernmental Panel on Climate Change has not even raised the question of global dimming in its official reports until quite recently. Likewise, whether global dimming is increasing or decreasing is also contested. Atsumu Ohmura's subsequent studies of satellite images of cloud coverage, for example, indicate that the skies may have brightened slightly since the early 1990s, and some these effects also have been observed elsewhere in other more focused studies. Yet, the complex interactions involved in global dimming make it difficult to use these studies in making reliable generalizations or long-term predictions.[18]

Although global dimming has been scientifically confirmed, its mitigation will be difficult, frustrating, and slow to attain. Given that its key catalysts are fossil fuels and/or chemical aerosol pollutants, humans cannot reduce global dimming unless cleaner forms of energy are found and fewer aerosols are developed. Even though the IPCC supported global dimming findings on a very limited scale in its 2007 reports, countermeasures—as the Montreal and Kyoto agreements have illustrated—can take years to negotiate. Once these countermeasures are hammered out, nations can and often do, flout their directives. What is more, reversing such contaminating environmental events could take a long time, or their effects might indeed prove irreversible, given that these declines can be documented at points all over the world.[19] From the 1950s to the 1990s, the level of solar radiation hitting various locales on the earth dropped significantly: 9 percent in Antarctica, 10 percent in the United States, 16 percent in the United Kingdom, 22 percent in Israel, and nearly 30 percent in parts of Russia. These new biophysics are yet to be reliably mapped, but preliminary surveys of what once was a predictable "nature" already show quite chaotic qualities in now much more unpredictable regional environments.[20]

Global dimming, then, is a complex and poorly understood process. As a *Science* article reported,

The climate of the Earth and its global mean surface temperature are the consequence of a balance between the amount of solar radiation absorbed by Earth's surface and atmosphere and the amount of long-wave radiation emitted by the system. The former is governed by the albedo (reflectivity) of the system, whereas the latter depends strongly on the atmospheric content of gases and particles (such as clouds and dust).[21]

Here the authors see the buildup of carbon dioxide and other greenhouse gases as promoting trends toward global warming. At the same time, however, greater aerosol concentrations and more clouds apparently cause enhanced levels of atmospheric albedo, or global dimming, which brings about global cooling effects.[22]

A Second Creation: From Nature to Urbanatura

The present trends, however, point toward anthropogenic sources of both greenhouse gases and aerosols as creating more pronounced atmospheric changes by 2050. Consequently, many experts claim that all of these observations "underscore the importance of understanding the natural and anthropogenic changes in Earth's albedo and the need for sustained, direct, and simultaneous observations of albedo with any methods that are currently available. Albedo changes may be as important as changes in greenhouse gases for determining changes in global climate."[23] Experts note that existing models for both global warming and dimming are limited by a lack of solid sensor and imaging data for the whole planet, so they argue in favor of being cautious in using both terms until more solid satellite-sensing data is available for detailed empirical analysis beyond the estimates provided by existing models.

Global warming, dimming, and/or cooling are the unintended consequences of human organisms reshaping the earth's natural and artificial environments to support their survival. And, as these moves are made, human and natural life forms begin to inhabit a nature that, as habitat, is being recreated by the output of corporate labs, major industries, and big agribusiness. Products and their by-products infiltrate terrestrial ecologies through human actions, and this technonature congeals in a "Second Creation," or urbanaturalized environments, with a new atmosphere, changing oceans, different biodiversity, and remade land masses.[24] Any study of climate change must consider all these ramifications.

Basically, an array of careful scientific observations from all around the world is providing strong evidence of tremendous anthropogenic alterations occurring in the atmosphere, and their effects cascading unexpectedly and unpredictably into radical changes in weather patterns, soil moisture levels, vegetation habitats, average sea levels, and terrestrial temperatures. Some believe that the droughts in the Sahel, heat waves in Europe, and more extreme weather patterns all around the planet can be connected to these rapid changes.

Here the deruralization of human communities, along with the denaturalization of the earth's environments, are combining into a more unpredictable, uninviting, and unpleasant hybrid of urbanism and nature, or an "urbanatura," for this and future generations to adapt their much more urbanized settlements to with few advance warnings and no obvious adaptive solutions. To even speak of "greenhouse gases" already implies the earth can now be best understood as an essentially built environment, a human-machine hybrid, or a vast artifice ironically fabricated by wastes, by-products, or effluents. Given that the causes of climate change are tied to burning fossil fuels for which there are few easily substituted alternatives, and granted that rapid changes could happen in a few days, weeks, or months, coping with the anthropogenic quirks of this urbanatura now challenges all tremendously.[25]

Global warming, dimming, and cooling have been building as serious threats to the earth's atmosphere for many decades, but it is the more recent profligate use of fossil fuels since the 1950s that has accelerated and concentrated earlier trends. Since the 1970s, the internationalization of industrial and agricultural production by ambitious global firms and aggressive national development agencies has also brought a neoliberal professional-technical worldview into ascendancy—one that helps cause but then denies these outcomes. It holds that

the world market eliminates or supplants political action—that is, the ideology of rule by the world market, the ideology of neoliberalism. It proceeds monocausally and economistically, reducing the multidimensionality of globalization to a single, economic dimension that is itself conceived in a linear fashion. If it mentions at all the other dimensions of globalization—ecology, culture, politics, civil society—it does so only by placing them under the sway of the world-market system.[26]

This globalist ideology pushes social forces to operationalize such beliefs and practices in a fashion that essentially requires states, societies, and economies to be managed like corporate capitalist enterprises, even though "this involves a veritable imperialism of economics, where companies demand the basic conditions under which they can optimize their goals."[27] Without a cohesive single state apparatus to oversee world society, globalist firms and elites enjoy, in turn, the most promising conditions possible for rapid growth since this "globally *disorganized* capitalism is continually spreading out, for there is no hegemonic power and no international regime either economic or political."[28] Whatever weak countervailing power exists on climate change mostly comes now from the IPCC and its more supportive nation-states using climatology as social critique.

With no central hegemonic force to restrain economic growth in world society, a ceaseless search for performance and profit on one level, as Lyotard claims, "continues to take place without leading to the realization of any of these dreams of emancipation."[29] Lacking any narratives of truth, enlightenment, or progress beyond using fossil fuels to gain growth for growth's sake, the scientific networks behind big business still push mass publics and markets to pursue more "goods" in economic growth by urbanaturalizing the planet.

Furthermore, global competition at this juncture is such that science often is compromised by commerce. As Lyotard asserts, governments and companies have foresaken "idealist and humanist narratives of legitimation in order to justify the new goal: in the discourse of today's financial backers of research, the only credible goal is power. Scientists, technicians, and instruments are purchased not to find truth, but to augment power."[30] Discovering that global dimming and warming exist as such reveals dangerous destruction, but it also presents new operational spaces in which all aspects of this transmogrified nature can be surveyed, reduced, and transformed in the scientific registers kept by contemporary capitalist companies and countries. Some scientists are becoming more critical, but many continue to pursue power and profit.

On another level, the "bad" by-products of these excessive levels of production growth products, such as global dimming, water shortages, soil erosion, weather disruptions, or biodiversity loss, could be negative indicators of "a new social system beyond classical capitalism," proliferating wildly across "the world space of multinational capital."[31] In fact, global warming, dimming, and cooling should be regarded as ecological markers of globalist flexible accumulation overshooting an array of highly contingent economic and ecological circumstances, like the earth's limited stores of fossil energy and atmospheric climate mechanisms, and "the result has been the production of fragmentation, insecurity, and ephemeral uneven development within a highly unified global space economy of capital flows."[32]

Omnipolitanization and Urbanatura

To some extent, globalist ideologies and transnational social forces now are combining to create an economy and society, which works at what Virilio terms an "omnipolitan" scale. Omnipolitanization represents both sides of the deruralization and hyperurbanization of the planet. Thus, the extreme concentration of commercialized values and economic practices in a *"world-city*, the city to end all cities," and, "in these basically eccentric or, if you like, *omnipolitan* conditions, the various social and cultural realities that still constitute a nation's wealth will soon give way to a sort of 'political' *stereo-reality* in which the interaction of exchanges will no longer look any different from the automatic interconnection of financial markets today."[33] Omnipolitanization brings globalizing neoliberal markets in "society" to conflate their imperatives with what were the material necessities of "nature," and the results are globalized artifices, like "urbanatura," unfolding behind, beneath, or beside the omnipolis. While other episodes of global warming, dimming, or cooling may have happened back in geological time, their current incidence appears only to have anthropogenic origins as well as dangerous implications for human communities worldwide. Global warming both is the most pervasive sign of this omnipolitan ambience, and the greatest pretext for making its omnipolitan order more concrete.

Consequently, serious analyses of culture, urbanism, and globalization today must recognize how omnipolitanizing tendencies are coevolving rapidly with the commodified ephemeralities of unthinking fossil fuel–driven global exchange. "Since movement creates the event," as Virilio asserts, "the real is *kinedramatic*."[34] The temperature increase of September 2001 after North American airline traffic was grounded is just one small piece of evidence here. Such kinedramatic global events flow through cohesive structures of production and consumption on a global scale, which become concretized in urbanatura as their petroleum-powered modes of operation hybridize the natural and the artificial in globalism's economic and social organization. It is these kinedramatic collectives of urbanatura that anchor the fragile ecologies and economies of the New World Order.[35]

In fact, urbanatura is a constructed world ecology/economy, whose spatiality and material have kinedramatic quiddity.[36] The massive energy grids that crisscross California, like a fragile electrical infrastructure that must deliver over 35,000 megawatts of electricity to power everyday life on a typical January day, do not also cover Colombia, Chad, or Cambodia, which use much less electricity every day in many of their rural cities and towns than one or two large office buildings alone require in downtown Los Angeles. The relations of the rich with the poor in urbanatura, particularly when using older concepts like "society" or "nature," cannot be explained well in solely nationalist, humanist, or realist terms.[37] Alternative terms of analysis, like urbanatura, must be found to reinterpret these relations, especially when California's multiple megawatts pollute the atmosphere, dimming and warming the globe for everyone, including those Colombians, Chadians, and Cambodians still struggling to produce and then use a few kilowatts more effectively so that they too might someday equal California's global dimming capabilities.

Global cooling, dimming, and warming are decisively significant ways in which a fossil fuel–burning, automobile-building, and commodity-buying culture has become naturalized in the geophysics of new weather patterns, soil conditions, and atmospheric conditions almost anywhere on the earth. Everyday life itself in the Group of 8 nations has a destructive ecological footprint that tracks through the earth's atmosphere, waters, climate, soils, and biodiversity.[38] As the IPCC reported in Shanghai during January 2001, "most of the global warming of the last 50 years," which could lead to average temperature increases of as much as 10.6 degrees, "is attributable to human activities."[39] Not all humans are equally responsible, but humans are apparently the main source of these drastic changes. The effluence of affluence is disrupting the earth's atmosphere as nature morphs into urbanatura; hence, it looks less and less like "Nature's Economy."[40] Environmental transfiguration is real, and much of it cannot be adequately addressed, much less effectively solved, without coming to terms with the mystified terms of resource depletion behind global exchange today.[41]

Nature as such, or the earth's environment before and/or apart from human activities, has not seen the current levels of CO_2 concentration, given that they have increased so rapidly over the past 250 years of the Industrial Revolution, in about 420,000 years.[42] Urbanatura, or the hybridities of humans' machinic metabolisms leaching into the earth's many ecologies with so many noxious products and by-products, constitutes an entirely new ecological order with its own energy flows, material exchanges, and habitat niches.[43] Global warming, dimming, and cooling are only the most evident atmospheric indicators of these changes. The United States, for example, is still barely 5 percent of the world's population, but its residents with all of their machinic infrastructures produce about a quarter of the earth's greenhouse gases—because they burn nearly 25 percent of all fossil fuel energy—as the global environments morph into urbanatura.[44]

On the one hand, the collectives of people and things in the United States are powerful enough to capture or control the production and use of sufficient oil, gas, and coal to generate massive energy inputs for their daily use. On the other hand, however, the inequality of these production and consumption linkages through the United States passes along its climate change by-products to scores of other nations. Their jointly produced exchanges offload the by-production of greenhouse gas emissions from limited ecological niches out across all of the other niches in the world's environment. Urbanatura, then, carves out its own numerous built-environmental niches where the modernization process ends, apparently leaving what constituted nature before global capitalism and the

Industrial Revolution behind for good.[45] Much of urbanatura appears now as hybridized material relations of inequality between highly urban deruralized countries, like the Group of 8 and other major OECD nations, and the more rural residents and refugees of other less wealthy and powerful economies and societies, like those in the Group of 77 countries.[46] The East African country of Uganda, for example, needs about 450 megawatts of electric power a day, but is now only producing about 100 megawatts, all from hydropower plants. Prolonged drought has drained Lake Victoria, so dams there only can generate less than a quarter of the country's needs. The results are daily blackouts, brownouts, or power outages. How much of this drought is normal, and how much of it is an outcome of global warming? Right now, no one knows. Climatology is addressing it, but could it be blamed on weather disruptions caused directly by the greenhouse gassing need to generate California's 35,000 megawatts of power?

Climatology must become social critique, because who and what cause these new worldwide environmental conditions, and who and what suffer from those hybridizing transformations, are becoming very serious questions. Even though many more nations are burning fossil fuels, using aerosol chemicals, and creating other atmospheric pollutants, climatology as social critique reveals how starkly material inequalities express themselves tangibly at every point of sale and site of production.[47]

Urbanature as a "Second Creation"

Nature is an essentially contested concept, and any study of global warming expresses its contested qualities. The centrality of a pure, objective, unmediated nature known accurately through the attainments of modern scientific knowledge is a notion that is dying very hard. With an ironic twist to Engels's famous characterization of socialism, the surge of surveillance data from satellites in space or sensors on the earth is moving many to think about forsaking the government of people to embrace the administration of things in urbanaturalized settings. In turn, these tools mediate new modes of control over people and things, which are expressed in many more partial, privatized, and productive practices in what is urbanatura.

From the vanguard of Newtonian physics in the seventeenth century to the rearguard of sociobiology in the twentieth century, many schools of modern science have assumed that their methodologies provide a privileged foundation for knowledge of what is "real" in nature as a definitive, methodologically rigorous mapping of a God-given creation that is truly "out there." These observations, in turn, are believed to reveal a true unsullied knowledge of objective reality of that Creation known now as "nature." This knowledge often is idealized in the mathematical proofs of physics, and its applications in everyday life are widely believed to be the foundations of modernity's technological proficiency. When all is said and done, humanity is believed to know how the worlds of nature function because of its disciplined application of scientific methods for observation, experiment, and verification. Yet, there now is more disquiet about these epistemological, ontological, and technological articles of faith in modernity.[48]

After the twentieth century, everyone must deal with postmodern conditions, which essentially are, as Jameson suggests, what prevail "when the modernization process is complete and Nature is gone for good. It has become a more fully human world than the older one, but one in which technoscientific products and by-products have become the basis of urbanatura's 'Second Creation.' "[49] Here the technical-economic conveniences of everyday life shake scientific technology's legitimacy, and trigger, at the same time, a reflexive realization that anthropogenic changes in the earth's climate, soils, atmosphere, waters, and biomass undercut incorrigible epistemic certainty about the planet's characteristics.

This new Second Creation is not as predictable as First Creation. On one level, the ecological opposition to modern science and technology is heartened by this realization, because their worries finally are registering in the theory and practice of contemporary scientists and technologists. Accordingly, new resistance movements argue that a more self-reflexive science could be less destructive of what was nature as well as more respectful of the human and nonhuman lives that still survive in the earth's many habitats. Yet, on another level, there are no guarantees for a positive outcome, because these individuals, along with everyone else who either openly supports or does not doubt modern science, find that whatever improvements in political power and economic property that millions have attained in the twentieth century depend to some degree on letting science continue to build on its technological proficiencies with new, exploitative operations in urbanatura. They need the goods and services made possible by the global economy's ongoing technical-economic productivity.

These beneficial outcomes are becoming more difficult to attain, however, because of the unpredictable effects of many industries' by-products and actual physical scarcities caused by resource depletion. Hence, urbanatura changes how the "environment" has been known:

The earth passes into the pure plane of immanence of a Being-thought, of a Nature-thought of infinite diagrammatic movements. Thinking consists in stretching out a plane of immanence that absorbs the earth (or rather, "adsorbs" it). Deterritorialization of such a plane does not preclude reterritorialization, but posits the creation of a future new earth. Nonetheless, absolute deterritorialization can only be thought according to certain still-to-be-determined relationships with relative deterritorializations that are not only cosmic but geographical, historical, and psychosocial.[50]

Whether it is variations in land topography, random differences in soil chemistry, water quality or weather, larger ecological pressures, land-use pressures, basic fishery overuse, general forest stress, or unpredictable atmospheric changes, urbanatura cannot be as readily surveyed or easily controlled an object of analysis as nature per se allegedly was. The reconstructed nature of urbanatura as a Second Creation, then, demands enveloping the earth in layers of bitspace for informatic surveillance, and tracking this data for material manipulation, simply to sustain most practices of agricultural and industrial production.[51]

Ultimately, these urbanatural transformations are a function of globalist restructurings in the world economy. If nature is gone for good, then the urbanatura of the Second Creation must be constantly monitored, measured, and then mastered at the organic and systemic level. Specializing in primary agricultural or forestry products is no longer necessarily a path to economic growth, or even stability for those already occupying those niches. Weird weather could wipe it all out in weeks. Consequently, new means of exploiting, or creating, comparative advantage in the global economy need to be discovered, and urbanatura requires new hybridized sciences and technologies to rerationalize transnational commerce at a national, regional, and local level. Only by seeking greater power and profit through high-tech sciences, for example, can comprehensive global accounts be kept of the planet's biomass to document humanity's apparent overdraft, sustainable abuse, or underutilization of these resources.[52]

From such systemic scans, the artificialized reconstruction of nature could be interpreted as a historical-geographic condition, a political-economic means of production, or a cultural-ethical regime of representation. All three of these possibilities reveal a unique spatial and temporal project that seeks to rewrite the codes of nature in the terms of urbanatural technification;

The *consumption* of individuals mediates the *productivity* of corporate capital; it becomes a productive force required by the functioning of the system itself, by its process of reproduction and survival. In other words, there are these kinds of needs because the system of corporate production needs them. And the needs invested by the individual consumer today are just as essential to the order of production as the capital invested by the capitalist entrepreneur and the labor power invested in the wage laborer. It is *all* capital.[53]

In this domain, ecology and economy merge as technoscience becomes both the key mode of production and the most embedded site of reproduction:

Everything has to be sacrificed to the principle that things must have an operational genesis. So far as production is concerned, it is no longer the Earth that produces, or labor that creates wealth . . . rather, it is Capital that *makes* the Earth and Labor *produce*. Work is no longer an action, it is an operation. Consumption no longer means the simple enjoyment of goods, it means having (someone) enjoy something—an operation modeled on, and keyed to, the differential range of sign-objects. Communication is a matter not of speaking but of making people speak. Information involves not knowledge but making people know.[54]

These maneuvers essentially write the new ontologues for urbanatura as a Second Creation of technifications in/of/through nature. Whether it is GIS-enabled biocomplexity modeling or a bioinformatically mapped transgenic organismic profiling, such reconstructions of nature are rendering, as Haraway claims, "thoroughly ambiguous the difference between natural and artificial, mind and body, self-developing and externally designed, and many other distinctions that used to apply to organisms and machines."[55] Recasting the world as machinic systems in order to surpass, but also acquire control over, the world as fungible matter and marketable code, is a project devoted to "systematizing something that is resolutely unsystematic, and historicizing something that is resolutely ahistorical,"[56] namely, exalting the imperatives of commodification through reconstructing nature 24/7. Indeed, the technified transformation of nature fulfills Haraway's anticipations of how contemporary ontologies must be propounded through "chimeras, theorized and fabricated hybrids of machine and organism."[57]

The best response to global climate change might not involve waiting for more extensive and intensive documentation of these trends. A more immediate and effective step would be to invoke the precautionary principle by acting as if these scientific findings are accurate, and moving aggressively to lessen the uses of fossil fuels and chemical aerosols forthwith. The unintended effects of these decisions could be as beneficial as the expected reduction of global warming or cooling inasmuch as they would entail finding more efficient means, as well as less unstable sources, for generating completely clean energy.

Amidst such uncertainty about nature, urbanatura seems more and more chaotic. Complex ecosystem dynamics, fragile atmospheric equilibria, and basic water, air, or soil chemistries have been compromised to the point that industrialization, citification, and deruralization processes are key formative forces at work in the earth's environment. It is not too late to invoke the precautionary principle to respond to these uncertainties, but it is becoming too late to count on such interventions to prevent serious, irreversible, or long-lived ecological degradation. After decades of environmental mobilization and resistance, the permafrost of Alaska is thawing, most coral reefs around the world are dying, and glaciers are retreating in the Alps, Greenland, the Andes, and British Columbia. Obviously, precautionary efforts plainly should be made, but our fragmentary knowledge of existing terrestrial ecologies as well as of their emerging anthropogenic disruptions will not necessarily ensure success by mitigating further harm on the margins of the planet's urbanaturalized ecologies.

As a complex site whose dimensions, directions, and determinations still remediate the extraction of surplus value for those seeking power and wealth, urbanatura entwines the political and economic in more intense systemic engagements of organic and inorganic interoperation. Hence, "the political," or those arrangements for who dominates whom, from the inside as well as from the outside of which governance systems, now must be examined. First and foremost, one must question this ecologized anthropogenic nature and all of its performative reconstructions, when pretending to speak about "the natural."[58] And, at the very least, these planetwide developments in urbanatura, like global climate change, challenge most of what has been understood to be a stable "environment," who "environmentalists" might be, and how those among them should work to "protect" this domain, since the Second Creation of urbanatura no longer should be regarded as what has been the subject of worry for traditional "environmentalism."[59]

Climatology as social criticism must move beyond today's ecological watchwords of *spatial* attention, or "think locally, act globally," into a new domain of *temporal* concern in which one must "think historically, act geologically" in the use of fossil fuels. If the climate carries the by-products of present and past combustion with all of their negative implications into the future long after the production and consumption of "goods" took place, then building up a store of environmental "bads" not observed for centuries, millennia or ages is unacceptable. As social critique, climatology then acquires an important new eschatological dimension by documenting how organic metabolisms for comfortable human life begin to undermine all humans' survival with a metabolic inorganicity capable of ending that same comfort. Air bubbles with carbon dioxide loads are being frozen in the Antarctic today at levels unequaled for 400,000 years—or 400 millennia. Such geological time frames should completely change how economic externalities, like industrial pollution or greenhouse gases, are regarded at any single point in time when they are generated. While protohuman stone tools have survived intact this long, virtually no good human products have lasted more than five millennia, but many bad human by-products ironically can last as long as 50, 80, or 100 times longer, in their climate-altering impact on the atmosphere, oceans, or soil of the earth itself.

Climatology and Capital's "Second Contradiction"

After over two centuries of such rapid, and radically inequitable, economic growth, many of the technical and organizational challenges of struggling against material necessity arguably have been met by the power of industrial technics. That is, with respect to the material cultures of advanced industrial society, any given set of "X" operational conditions for many industries allows them to produce and distribute virtually any range of "Y" products. Yet, is there also a less apparent, and more insidious set of "Z" by-products that go along with these celebrated instrumentally rational efficiencies? And is environment destruction one of the most serious instances of this more stealthy set of unassessed externalities? If sustainable development exists, then it is, in fact, an order of mystification rooted in schemes for legitimating systemic degradation as outcome "Z" begins to undercut the initial conditions of "X" to produce outcome "Y."

This system of sustainable degradation implicitly concedes, or explicitly extends, as it cynically builds on, the "second contradiction" of capitalism as identified by O'Connor.[60] Acknowledging that the underproduction of capital coupled with the destruction of nature becomes a means of producing knowledge about this new economic environment as well as an opportunity for mobilizing powers to cope with its environmental effects, O'Connor suggests how systemic ecological degradation is never halted. It is instead measured, monitored, and manipulated within given tolerances as ecological devastation is sustained in the contradictory urbanatura of capitalist-built environments.

Despite intense efforts by scores of movements in dozens of countries over many decades, the capitalist modes of production, consumption, accumulation, and circulation behind greenhouse gassing persist. Indeed, the adaptability of this polyglot and fluid economic formation, while being indicted by climatology for climate destruction, has proven quite remarkable. A grassroots desire for ecological sustainability is real, but its articulation under existing juridicolegal conditions of governance necessitates that those energies be captured, contained and then channeled into more commodified options and conventional practices within today's operational parameters for global capitalism. Climatology as social criticism is just one example of these adaptive responses.

Some still might struggle to develop a genuinely sustainable ecological society, but most have seized on, or surrendered to, the business opportunities created by today's popular sustainability rhetorics to instead emphasize market-based development. After decades of increasing greenhouse gas emissions, it appears that sustainable development, partly by accident and partly by design, is, in fact, a system of sustainable degradation to manage humanity's destructive interactions with nature. These strategies become collective solutions, which are forged in reaction to contemporary global capitalism's "capital underproduction and unproductive use of capital produced."[61]

During the 1980s and 1990s, the continuing crises of capitalist production necessitated massive restructurings, and these changes continue in the 2000s. As O'Connor has observed, the continuous reorganization of capital must rely on the increasing "variability of the use of labor power, flexibility of other forms of capital, and cuts in wage and other production costs, on the one hand, and a dangerous expansion of externalities or social costs and a shameful neglect of production conditions, on the other."[62] Environmental disrepair is, of course, being constantly discovered to be in dire need of mitigation. The orchestrated awareness by the IPCC of global climate change is important, but the IPCC's experts are only providing superficial remedies, here and there, for the environment's degradation.

Global climate change is a clear sign of severely stressed "conditions of capitalist production."[63] O'Connor identifies them, first, as "external physical conditions," or natural elements required for constant and variable capital; second, as the "labor power" of workers entangled in their own personal conditions of production; and third, as the "communal, general conditions of social production."[64] He argues that contemporary ecological critiques should take a far more expansive look at the overall articulation and regulation of "the conditions of production." Of course, the IPCC cannot easily do all of this work. Consequently, he asserts,

Today "external physical conditions" are discussed in terms of the viability of eco-systems, the adequacy of atmospheric ozone levels, the stability of coastlines and watersheds; soil, air and water quality; and so on. "Laborpower" is discussed in terms of the physical and mental well-being of workers; the kind and degree of socialization; toxicity of work relations and the workers' ability to cope; and human beings as social productive forces and biological organisms generally. "Communal conditions" are discussed in terms of "social capital," "infrastructure," and so on. Implied in the concepts of "external physical conditions," "laborpower," and "communal conditions" are the concepts of space and "social environment," which in turn [sic] helps to produce social environments. In short, production conditions include commodified or capitalized materiality and sociality excluding commodity production, distribution, and exchange themselves.[65]

While the dependence on continuous crisis in contemporary capitalism necessitates changes in overall productive forces and forced changes in the social relations affecting productive conditions, the IPCC vets the most suitable climate change schemes that might help advance control, guide planning, and express flexibility under these operational conditions. Nonetheless, the global market still continues to bring more exciting products to life for wealthy consumers, and it still leaves the more toxic by-products for poor producers.

Social criticism of climate-related issues today by many scientists, ironically, is not unlike critical thinking about commerce in the nineteenth century among many socialists. Crisis is their shared register, and both talk of an intrinsic, unalterable, and inexorable set of tendencies that threaten everyday life as we know it. Purposeful rational action at the household, and more importantly the company, level is leading to increasingly intense and unstable anarchic relations at the national and global level of action. Rapid quantitative increases in the economy's products as well as its by-products are leading possibly to a qualitatively different order, which once attained cannot be easily reversed. Embedded automatic imperatives seem to move events ahead in a predictable fashion, but no one single predictive framework holds enough sway in any one society to create consensus over how to proceed.

The abuse of shared atmospheric resources is clearly leading to another tragedy of the commons, but the commonness of this tragedy prevents effective efforts at making significant changes in the existing institutions that should be creating these changes. Like socialism as a discourse of social critique, climatology is also fracturing into many different varieties of critical theorizing in desperate struggles to try to find some traction politically. Frustrated over the consumers' and producers' unwillingness to forsake fossil fuels to forestall global warming, some climatologists already have assumed the stance of a vanguard group of more conscious, committed, and collective cadres ready to push their own ecofuturist alternatives based on IPCC edicts about fossil fuel use, anticipatory climate management, and ordinary democratic deliberation.

Shocked by mass inattention, government inaction, and corporate inefficiency, IPCC-friendly climatologists (or perhaps IPCCrats?) mobilize their limited, but nonetheless real, global authority to call attention to how the world's economies and societies should be administered amidst this perversely rapid transition into pervasive global warming. While their frustrations are real, it is unclear that many want, or even understand, what this IPCC-rooted governmentality implies for mass publics living under such IPCCratic authority. Nonetheless, a powerful push for atmospheric governmentality is building on the coming crisis narratives in the earth's environment from critical climatology, which is blazing a tiny trail for IPCCarchy.

In 1988, O'Connor envisioned such crisis-induced restructuring as a chance to exercise more control by capital through collective planning. Whether corporate or state, new forms of flexible planning and planned flexibility indeed have been created, as the IPCC illustrates. The Kyoto Protocol is a significant achievement. And it was, after all, negotiated and implemented by global business, technoscience, and governance organizations, even though it bears fatal flaws. Still, this cluster of global policy responses is also now so integrated into the system of sustainable degradation that one can ask if it only ideologically clads it with a patina of ecojuridical propriety rather than serving as a strategy for positive ecological transformation.

Thinking about O'Connor's second contradiction, then, it is clear that the IPCC's manifold efforts to reduce greenhouse gases are insufficient to manage, mitigate, or, if needed, manipulate the damage inflicted on nature as a spatial domain to be controlled as entire ecosystems, biomes, or environments. The social critique of climatology is real, but can it ever get past accepting sustainable degradation? Climatologists admit there indeed is a crisis, and then they seek to respond in a proactive, profitable, and powerful fashion. Yet, does the work of the IPCC only mask negative outcomes, maintain some environmental viability, and create zones of control where degradation is at best lessened, but greenhouse gassing is never stopped? The growing number of scientific studies heighten awareness of climate change, yet it is rarely stemmed. The existing inequality of commodity production and consumption spills over into new inequalities in commodity by-production and consumer choicelessness, because technoscience is left only to scrupulously document additional biospheric losses. However, it cannot easily change how loss is incurred.

As O'Connor maintains, such changes "either typically presuppose or require new forms of cooperation between and within capitals and/or between capital and the state and/or with the state, or more social forms of the 'regulation' of the metabolism between humankind and nature" as well as the " 'metabolism' between the individual and the physical and social environment."[66] Climatology as social critique only accentuates corporate cooperation; and, thus, one finds "more cooperation has the effect of making production conditions (already politicized) more transparently political, thereby subverting further the apparent 'naturalness' of capital's existence."[67]

The system of sustainable degradation ensures that limited democratic challenges to corporate roles and responsibilities will be launched by exercising the prerogatives of technoscientific expertise in global forums meant to monitor the workings of liberal democratic societies. Nonetheless, corporate expertise and private property constitute the key material forms of real power within the existing conditions of production. With most businesses and professions tied to greenhouse gas emissions, the experts and owners are treated by common practices, unspoken assumptions, and conventional laws as distinct centers of greater authority with a more special legitimacy than IPCC-linked climatology. It is this sort of narrowly construed and questionably legitimated power that liberal democratic capitalism opens to the public, and many social movements have contested it in public settings over the past couple of centuries of capital's development with a mixed record of success. Critical climatology continues this tradition, but it too succeeds only within very narrow limits.

Climatology as critical theory ultimately turns its historical meteorology into an applied policy discourse about ecofuturism. Given the complexity of so many long-run trends, the unpredictability of so many immense changes, and the uncertainty over whether any preparations for the worst or hopes for the best will pan out, climatology has innumerable opportunities to hold forth about "what must be done?" Greenhouse gassing continues unabated. And, because efforts to check it are caught in gridlock, some critical climatological discourse veers into making indefensible normative claims. While admitting that climatological science is hazy, climatologists as critical theorists do assert that nature "is" characterized by certain predictable features and that, on the basis of these predictions about global warming, weather disruption, rising seas, and ocean changes, governments and firms "should be" doing A, B, and C to respond.

Yet, these ecofuturistic policy recommendations do not have a compelling quality to them. Even though scientists try to derive normative oughts from predictions about positive "is-ness," such efforts lack logical necessity. Moreover, the contestedness of their positive "is-ness" readings undercuts the solidity of the "ought-ness" recommended as solutions to the challenges at hand. This handwringing style of ecofuturist policy discourse creates the illusion of understanding the threats, the appearance of responding rationally to the pressures rising in the environment, and the image of geophysical science guiding society toward solutions to serious economic and security problems as the climate changes. Unfortunately, most ameliorative responses thus far are taking place only on the margins of change. Critical thinking about climate change is still contested in the policy and science communities, and the ecofuturist narrative of flooded nations, eroding coasts, superdestructive storms, and vicious weather has not triggered the total reconstitution of contemporary capitalist societies and economies to reduce greenhouse gases to less threatening levels. If anything, the embedded path dependencies of fossil fuel combustion that began during the eighteenth and nineteenth centuries have only hardened and deepened, leaving the dire ecofuturist critique of climate change to become a virtually self-fulfilling prophecy.

Nonetheless, deploying climatology as social critique is still organizationally important. Such analysis has made "capital and state" confront some of their basic contradictions, which then can be displaced into these "political and ideological spheres (twice removed from direct production and circulation)," while, at the same time, "there is introduced more social forms of production. Conditions defined both materially and socially, e.g., the dominance of political bipartisanship in relation to urban redevelopment, educational reform, environmental planning, and other forms of provision of production conditions which exemplify new and significant forms of class compromise."[68] IPCCarchic authority propounds itself into technoscientific coalitions, and new international compromises might extend today's weak agreements about incremental reductions in greenhouse gassing, but there is no necessarily imminent form of socialism congealing here. For some, IPCC actions might serve as a step toward making socialism at least more imaginable, but in the meantime they are seen as a small spanner in the works of global exchange.

Of course, such small spanners in the works often soon become simply another buttress of the works, because critical climatology can also morph into another applied variant of environmentality. Although not yet as visible as marine experts behind the "Law of the Sea," some vanguard climatologists already are nominating themselves and their networks to help draft the "Law of the Atmosphere." Believing that climatological science as social critique is ready for the policy area, some academic institutions, like Columbia University in New York, are bundling together programs of climate science, earth science, and social science to influence decision making. As Columbia's new MA program in Climate and Society suggests,

Recent research has generated a wealth of new knowledge about long-term climate change, shorter-term climate variability, and their socioeconomic impacts. Decision makers need clear and reliable guidance on impending climate shocks, as well as practical information and tools to deal with their consequences. Building on improved scientific understanding of climate and improved coping mechanisms, Columbia University is training a new generation of academics and professionals at the nexus of social science, climate science, and public policy.[69]

While admitting that global warming is still a contested proposition, and either anthropogenic or nonanthropogenic forces might be behind it, Columbia's environmentality-minded experts are content to train new climatological careerists capable of claiming power through knowledge: "Regardless of the outcome of this scientific debate, there is an emerging consensus in governments, corporations, and civil societies that climate anomalies damage societies and need to be factored into policy decisions and economic strategies."[70]

Determining Ends

To conclude, this analysis of global warming, dimming, and cooling as social creations and social constructions returns to political economy. The mobilization of climatology as social critique constitutes one of the more salient intellectual and institutional strategies used today in answer to the second contradiction of capital identified by O'Connor. And climatology as social criticism is finding social institutions to adapt the destructive global conditions of economic production to today's environmental crises, even as the public and policymakers acknowledge the material reality of global warming, global dimming, or global cooling. Without a material critique of political economy, however, climatology is not enough. IPCCarchic deliberations permit those in power to draw upon capitalist culture itself in order to say "something is being done" in response to climate change, but what is being done might amount to little more than the systematization of ecological degradation in economically sustainable forms. Getting MA degrees in Climate and Society is just one bad sign of such things to come. In the cynical calculus of capital, much value is to be found in overstretching the earth's resources; what is more, the system of sustainable degradation enables capital to extract even more value by maintaining the appearance of sustainability while exploiting more finely the raw realities of nature's degradation.

Climatology as social criticism maps how the unintended consequences of industrial capitalism are externalized as by-products of mass production and consumption, only to begin altering the earth's atmosphere. At one time, "scientific socialism" presumed to foretell the workers of the world about the coming crisis of capitalism, out of which would come a more rational, just, and equitable communist order. An intrinsic set of tendencies were believed to be creating the basis for full rationalization of the means of production as well as the opportunity to enact new forms of material equality, political deliberation, and psychosocial emancipation. Unalterable laws of surplus value would guarantee the advent and permanence of these outcomes as the chaotic dynamics of the market pushed the anarchy of exchange toward the order of communism.

While its scientific credibility clearly exceeds that of historical materialism, contemporary climatology, especially in its more engagé expressions as public policy, popular science, or economic forecasting, often strangely echoes, parallels, or reimagines postulates not unlike those of the materialist conception of history. While it clearly is not completely the same, it also is not entirely different. These complementarities, convergences, or commonalities, in turn, deserve closer attention. The circulation of climatological analyses focused on global warming ultimately shows the poverty of prognostication. Good science with reliable findings about global warming trends has been available to decision makers for decades. Yet, during these same decades, very little has been done effectively to reduce net greenhouse gas emissions beyond identifying and aiming at a future ceiling level pegged to floor values measured in 1990. Valid scientific studies with reliable predictive findings, in fact, have been consistently downplayed, derided, or dismissed by corporate and government leaders.

Nonetheless, critical climatology persists. Climatologists already are pulling these threads together, moving from social critique to training "ecomanagerialists."[71] As the director of Columbia University's Climate and Society MA program asserts,

Advances in climate modeling and prediction have changed the landscape of human knowledge. For drought-stricken farmers of the developing world, for shantytown dwellers at the mercy of hurricanes and mud slides, for governments trying to make the most of limited resources as they strive for development, and for the multibillion dollar insurance and food industries, this new scientific knowledge can offer better ways to respond to the problems and opportunities created by a varying climate. But decision makers must understand how to make effective use of this new knowledge.

The need for professionals who understand the links between climate and society is acute, and grows ever more so as human activity alters the global atmosphere. The Columbia M.A. in Climate and Society will give you the knowledge and skills to meet this need.

Columbia University is home to leading researchers in the fields of climate change, climate prediction, and earth and atmospheric sciences. We have unparalleled experience in training policymakers, leaders, and thinkers in the heart of New York City, home to the United Nations and the seat of world politics.

One innovative program brings together dedicated individuals from all over the world to study and shape our common fate.[72]

Even though this scientific knowledge about the atmosphere is limited and unreliable, some clearly see opportunities to leverage what little is known against all that is unknown to operate near the "seat of world politics."

Even so, their ability to act is highly constrained by the politics of fossil fuel dependence. Despite all of the lip service given to the centrality of scientific calculation, predictive positivist modeling, and instrumental rationality, the prognostications of climatology as good science are still essentially ignored. At best, their effects are felt on the margins of change in scattered municipal-, regional-, or provincial-level policies, occasional directives from smaller, less polluting states, or individual efforts at conservation. Such shifts are salutary, yet they are not anywhere near as foundational, permanent, or thoroughgoing as they should be in order to effect a fundamental reversal of the greenhouse gas problem.

Of course, the real problem here is that science itself on its own actually provides no authoritative, fail-safe, or precise guidance about what ought to be done in response to global climate change. Indeed, it only continues to chronicle what is being done, while hypothesizing about how the effects of what is being done could affect the earth's climate. The inertia of centuries past in greenhouse gas emissions makes it unlikely that immediate drastic changes would necessarily improve global warming trends. Things should get better, but it is not known precisely how soon, to what extent, or with what side effects. The real uncertainty over anthropogenic and nonanthropogenic causes as well as their current interactive effects also lessens the authority of scientific assessments. Scientific neutrality must acknowledge the possibility that current global warming could be a relatively periodic, if not cyclical, occurrence stemming from variations in solar activity, ocean current circulation, natural greenhouse gassing, or other yet to be determined nonanthropogenic causes. Human-made greenhouse gases are not improving these dynamics, but it is still unclear what relative importance can be given to anthropogenic versus nonanthropogenic sources of greenhouse gases. As too many experts still conclude, the evidence can be seen as mixed.

Global warming, as a scientific phenomenon as well as a public policy problematic, also possibly represents an instance of ideological mystification even as different schools of analysis decry its deleterious effects. The concentration of anthropogenic sources of greenhouse gases has created a trend of such magnitude and duration that it essentially is naturalized. When urbanatura displaces nature, quibbling over anthropogenic versus nonanthropogenic sources could be an accounting error. Because "greenhouse gassers" basically are each and every individual who engages directly or indirectly in the combustion of hydrocarbon fuels, the greenhouse effect also becomes a collective expression of so many individual decisions and nondecisions. In this manner, there often is an erasure of casual agency. Mass markets of individual consumers, who create these industrial by-products, become reified into entities like "greenhouse gas emissions," whose origins and operations also are occluded by the machinic metabolisms of global production and consumption.

When diced and sliced into conventional, but not always instructive, "national" accounts of greenhouse gas producers, the tallies may well be accurate, but also not true to reality. The indefinite nature of responsible agency dissipates into too many anonymous masses. The system of technology-consuming fossil fuels, and a more obtuse and obdurate structure of definite gassing trends, spins up from natural input-output tables of oil, gas, and coal consumption in hybrid machinic metabolisms. Clearly, more astute accountancy of ecological footprints could trace some of these trends back to the household or individual levels of use, but most of these calculations are simplistic per capita divisions of gross natural products. Figures can be calculated, but no one can figure out which figures—individual or collective—to blame.

Moreover, the accurate indictment of such personal environmental wrongs loses political traction as soon as "everyone" becomes responsible. Anyone therefore is accountable, but no one then can do much to change the trends for which they are all held accountable. Hence, there is a grudging acceptance of capitalism's destructive action, because it all too often must be excused as inevitable, unavoidable, or unanticipated as second- and third-order implications of so many billions of initial decisions to burn more carbon-releasing matter at all fossil fuels' individual points of purchase and use. If indefinite productive collectives cause the crisis, then advocating corrections that mystify their structural imperatives to produce by pinning full responsibility on the agency of definite individual consumers does little to alter crucial structures.

Indeed, one will be left with energy-producing transnational firms, like British Petroleum, telling consumers to choose to live "beyond petroleum" when producers now have no real good energy alternative behind their fossil fuel products to live "beyond the profits" created when pretending BP and its customers truly can move, in this instance, "beyond petroleum" or, more importantly, "British Petropower." Such farces about collective responses to climate change call into question the liberal order allegedly based on social contracts drawn up in the state of nature. Indeed, they highlight the tragic necessity of renegotiating most social relations—contractual and otherwise—to fit today's states of urbanatura, and in ways yet to be determined, both ethically and politically.

Notes

Passages from this chapter are drawn from *Rethinking Globalism*, ed. Manfred Steger (Lanham, MD: Rowman & Littlefield, 2004); *Capitalism Nature Socialism*, 16 (2005); and *Alternative Globalizations: 2006 Conference Documents*, ed. Jerry Harris (Chicago: LuLu.com, 2006).

1. See http://www.globalissues.org/EnvIssues/GlobalWarming/Globaldimming. ASP for an account of these two contradictory effects, which are both caused by using fossil fuels that leave greenhouse gases and particulate pollutants. These declines parallel the increase of CO_2 and other pollutants into the atmosphere since the seventeenth century. At that time, each human inhaled 280 molecules of CO_2 with each breath, but today every human inhales 380 CO_2 molecules

with each breath. And this level increases about two molecules a year. See Robert H. Socolow, "Can We Bury Global Warming?", *Scientific American* 293, no. 1 (July 2005): 49.

2. Douglas Long, *Global Warming* (New York: Facts on File, 2004), 60.

3. Timothy W. Luke, "Liberal Society and Cyborg Subjectivity: The Politics of Environments, Bodies, and Nature," *Alternatives: A Journal of World Policy* 21, no. 1 (1996): 1–30.

4. Long, *Global Warming*, 61–63.

5. Long, *Global Warming*, 61–63.

6. Mark Maslin, *Global Warming* (Oxford: Oxford University Press, 2004), 4.

7. Maslin, *Global Warming*, 14.

8. Long, *Global Warming*, 4.

9. Bjørn Lomborg, *The Skeptical Environmentalist: Measuring the Real State of the World* (Cambridge: Cambridge University Press, 2001), 261–263.

10. Lomborg, *The Skeptical Environmentalist*, 261.

11. William K. Stevens, *The Change in the Weather: People, Weather, and the Science of Climate* (New York: Delta, 1999), 18.

12. Maslin, *Global Warming*, 52.

13. Maslin, *Global Warming*, 55.

14. Maslin, *Global Warming*, 73–74.

15. See Gerald Stanhill and Shabtai Cohen, "Global Dimming: A Review of the Evidence," *Agricultural and Forest Meteorology* 107 (2001): 255–278. Fossil fuels and their greenhouse gas emissions, like CO_2, are the main source of these problems. From 1751 to 2002, it is estimated by the International Energy Agency that around 1,070 billion tons of CO_2 were emitted into the atmosphere. Of this total, 542 billion tons came from coal, 142 billion from gas, and 386 billion from oil; moreover, the next generation alone is predicted to be almost as dirty, because 501 billion tons from coal, 226 million from gas, and 8 billion from oil for a total of 735 billion tons will have been vented into the air between 2003 and 2030 (Socolow, "Can We Bury Global Warming," 52).

16. See http://www.bbc.co.uk/sn/tvradio/programmes/horizon/dimming_qa.shtml for an account of this work by David Travis at the University of Wisconsin.

17. For more insight into Ohmura's scientific work, see David Adam, "Goodbye Sunshine," *The Guardian*, December 18, 2003. Also see A. Ohmura, "Reevaluation and Monitoring of the Global Energy Balance," in M. Sanderson, ed., *UNESCO Source Book in Climatology*, 35–42 (New York: UNESCO,: 1990).

18. See Peter Christhoff, "Weird Weather and Climate Culture Wars," *ARENA Journal* 23 (2005): 9–17. The complexity of such climatic change, as Christhoff observes, is now grappled with more broadly in works of popular fiction and film, like Michael Crichton's novel *State of Fear* or the film *The Day After Tomorrow*.

19. See Michael Roderick and Gerald Farquhar, "The Cause of Decreased Pan Evaporation over the Past 50 Years," *Science* 298 (2002): 1410–1411. These observations were simple but strong. Scientists have monitored a considerable

drop in the pan evaporation rate all over the world, which simply amounts to measuring how much water evaporates from pans each day. These elementary measures have been taken since the nineteenth century, so the data are extensive and well documented. On average, most pans had about 100 millimeters less water evaporate in the last three decades. Since each millimeter takes about 2.5 megajoules of solar energy, 100 millimeters less in thirty years in Russia, for example, means a decline in sunlight of 250 megajoules. This figure, in turn, matched measurements taken in the United States and Europe. See http://www .bbc.co.uk/sn/tvradio/programmes/horizon/dimming_trans.shtml.

20. See Beate Lepert, "Observed Reductions in Surface Solar Radiation at Sites in the U.S. and Worldwide," *Geophysical Research Letters* 29, no. 10 (2002): 1421–1433. Indeed, it is radical unpredictability and extreme phenomena that make the visions of *State of Fear* and *The Day After Tomorrow* so compelling as glimpses of a new kind of "nature."

21. Robert J. Charlson, Francisco P. J. Valero, and John H. Seinfeld, "Atmospheric Science: In Search of Balance," *Science* 308, no. 5728 (May 6, 2005): 806–807.

22. M. Wild, H. Gilgen, A. Roesch, A. Ohmura, C. Long, E. Dutton, B. Forgan, A. Kallis, V. Russak, and A. Tsvetkov, "From Dimming to Brightening: Decadal Changes in Solar Radiation at Earth's Surface," *Science* 308 (May 6, 2005): 847–848.

23. Charlson, Valero, and Seinfeld, "Atmospheric Science," 806.

24. See Timothy W. Luke, "Reconstructing Nature: How the New Informatics Are Rewrighting Place, Power, and Property as Bitspace," *Capitalism, Nature, Socialism* 12, no. 3 (September 2001): 87–113.

25. See J. Chen and A. Ohmura, "Estimation of Alpine Glacier Water Resources and Their Change Since 1870s," *Hydrology in Mountainous Regions 1, IAHS Publication* 193 (1990): 127–135.

26. Ulrich Beck, *What Is Globalization?* (Oxford: Blackwell, 2000), 9.

27. *What Is Globalization?*, 9.

28. *What Is Globalization?*, 16.

29. Jean-François Lyotard, *The Postmodern Condition: A Report on Knowledge* (Minneapolis: University of Minnesota Press, 1984), 39.

30. Lyotard, *The Postmodern Condition*, 46.

31. Fredric Jameson, *Postmodernism, or the Cultural Logic of Late Capitalism* (Durham, NC: Duke University Press, 1991), 59, 54.

32. David Harvey, *The Condition of Postmodernity* (Oxford: Blackwell, 1989), 296.

33. Paul Virilio, *Open Sky* (London: Verso, 1997), 75.

34. Paul Virilio, *The Art of the Motor* (Minneapolis: University of Minnesota Press, 1995), 23.

35. William Greider, *One World, Ready or Not: The Manic Logic of Global Capitalism* (New York: Simon & Schuster, 1997), 11–53.

36. See Timothy W. Luke, "At the End of Nature: Cyborgs, Humachines, and Environments in Postmodernity," *Environment and Planning A*, 29 (1997): 1367–1380. The omnipolitan sweep of urbanatura gives it planetary policing and protecting problems. Indeed, this fusion of urbanism with environmental disruptions is typified by plans to "capture," or "sequester," or "sink" CO_2 remissions from fossil fuels within pollutant prisons in old oil fields, in deep rock formations, or even at the bottom of oceans to keep it from entering the atmosphere to warm and/or dim it. Here urbanatura is envisioned as an integral production, circulation, consumption, and accumulation mechanism to control the greenhouse gases that cause urbanatural blight. See Howard Herzog, Balour Eliasson, and Olav Kaarstad, "Capturing Greenhouse Gases, *Scientific American* 282, no. 2 (February 2000): 72–79; S. Pacala and Robert Socolow, "Stabilization Wedges: Solving the Climate Problem for the Next 37 Years with Current Technologies," *Science* 305 (August 13, 2004): 968–972; Soren Anderson and Richard Newell, "Prospects for Carbon Capture and Storage Technologies," *Annual Review of Environment and Resources* 29 (2004): 109–142.

37. Lewis Mumford, *The City in History: Its Origins, Its Transformations, and Its Prospects* (New York: Harcourt, Brace & World, 1961); J. R. McNeill, *Something New under the Sun: An Environmental History of the Twentieth-Century World* (New York: Norton, 2000).

38. Leon D. Rotstayn and Ulrike Lothmann, "Observed Reductions in Surface Solar Radiation at Sites in the U.S. and Worldwide," *Geophysical Research Letters* 29, no. 10 (2002): 1421–1433; Real Climate, "Global Dimming" (January 18, 2005), http://www.realclimate.org/index.dhd?p=los.

39. Philip P. Pan, "Scientists Issue Dire Prediction on Warming," *Washington Post*, January 23, 2001, A1.

40. Donald Worster, *Nature's Economy: The Roots of Ecology* (Garden City, NY: Anchor Books, 1979).

41. Hilary F. French, *Vanishing Borders: Protecting the Planet in the Age of Globalization* (New York: Norton, 2000).

42. Pan, "Scientists Issue Dire Prediction on Warming," A1.

43. See Timothy W. Luke, "Cyborg Enchantments: Commodity Fetishism and Human/Machine Interactions," *Strategies* 13, no. 1 (2000): 39–62; Luke, "Liberal Society and Cyborg Subjectivity."

44. Vaclav Smil, *Energy in World History* (Boulder, CO: Westview Press, 1994).

45. Michael J. Dear, *The Postmodern Urban Condition* (Oxford: Blackwell, 2000); Edward Soja, *Postmetropolis: Critical Studies of Cities* (Oxford: Blackwell, 2000); Michael Peter Smith, *Transnational Urbanism: Locating Globalization* (Oxford: Blackwell, 2000).

46. Robert D. Kaplan, *The Ends of the Earth: A Journey at the Dawn of the 21st Century* (New York: Random House, 1996).

47. Timothy W. Luke, "Placing Powers, Siting Spaces: The Politics of Global and Local in the New World Order," *Environment and Planning A: Society and Space* 12 (1994): 613–628.

48. See Bruno Latour, *The Politics of Nature* (Cambridge: Harvard University Press, 2004).

49. Jameson, *Postmodernism, or the Cultural Logic of Late Capitalism,* ix.

50. Gilles Deleuze, *What Is Philosophy?* (New York: Columbia University Press, 1994).

51. See Luke, "Reconstructing Nature," 110–113.

52. Luke, "At the End of Nature."

53. Jean Baudrillard, *For a Critique of the Political Economy of the Sign* (St. Louis, MO: Telos Press, 1981), 82.

54. Jean Baudrillard, *The Transparency of Evil: Essays on Extreme Phenomena* (London: Verso, 1993), 45–46.

55. Donna Haraway, *Simians, Cyborgs, and Women* (New York: Routledge, 1991), 152.

56. Jameson, *Postmodernism, or the Cultural Logic of Late Capitalism,* 418.

57. Haraway, *Simians, Cyborgs, and Women,* 150.

58. See Latour, *The Politics of Nature.*

59. See Timothy W. Luke, "Global Cities vs. global cities: Rethinking Contemporary Urbanism as Public Ecology," *Studies in Political Economy* 71 (2003): 11–22.

60. James O'Connor, "Capitalism, Nature, Socialism: A Theoretical Introduction," *Capitalism, Nature, Socialism* 1 (fall 1988): 11–39.

61. O'Connor, "Capitalism, Nature, Socialism," 27.

62. O'Connor, "Capitalism, Nature, Socialism," 7.

63. O'Connor, "Capitalism, Nature, Socialism," 16.

64. O'Connor, "Capitalism, Nature, Socialism," 16.

65. O'Connor, "Capitalism, Nature, Socialism," 17.

66. O'Connor, "Capitalism, Nature, Socialism," 27.

67. O'Connor, "Capitalism, Nature, Socialism," 27–28.

68. O'Connor, "Capitalism, Nature, Socialism," 29–30.

69. See http://www.columbia.edu/cu/climatesociety/aboutclimate.htm.

70. See http://www.columbia.edu/cu/climatesociety/aboutclimate.htm.

71. See Timothy W. Luke, "Training Eco-Managerialists: Academic Environmental Studies as a Power/Knowledge Formation," in Frank Fischer and Maarten Hajer, eds., *Living with Nature: Environmental Discourse as Cultural Politics,* 103–120 (Oxford: Oxford University Press, 1999); Timothy W. Luke, "Environmentality as Green Governmentality," in Eric Darier, ed., *Discourses of the Environment,* 121–151 (Oxford: Blackwell, 1999).

72. See http://www.columbia.edu/cu/climatesociety/director.html.

7

Urban Sprawl, Climate Change, Oil Depletion, and Eco-Marxism

George A. Gonzalez

The U.S. urban zones are the most sprawled in the world.[1] In the context of spiking oil prices and the contribution of urban sprawl to the global warming or climate change phenomenon, urban sprawl in the United States has been garnering attention.[2] Urban sprawl can only be fully comprehended within the political economy framework developed by Karl Marx. Marx's concepts of value and rent are indispensable to understanding the profligate use of fossil fuels—vis-à-vis urban sprawl—that has significantly contributed to oil depletion[3] and to the recent global warming trend. This argument is consistent with the eco-Marxist contention that the writings of Marx and Frederick Engels contain a thorough ecological critique of capitalism.[4]

Urban sprawl was deployed in the United States during the 1930s as a means of reviving U.S. capitalism from the Great Depression. The sprawling of urban zones greatly increased the need for automobiles and other consumer durables. This use of urban sprawl to increase economic demand is consistent with Marx's argument that demand within capitalism is malleable and is geared toward increasing the consumption of goods and services produced through social labor.[5] The exploitation of social labor is the basis of capitalist wealth.[6]

The development of pro–urban sprawl policies in the United States is also consistent with the business-dominance view of public policy formulation. Those that hold this view of the policymaking process contend that economic elites and producer groups are at the center of public policy formation.[7]

Overview

I begin by explaining that Marx's concept of exchange value results in the idea that raw materials have zero exchange value within capitalism.

I then use the U.S. petroleum market to demonstrate the validity of this supposition. It is precisely because petroleum has no intrinsic exchange value and because of the historically abundant supplies of oil in the United States[8] that urban sprawl could be utilized in the United States beginning in the 1930s as a means to absorb the productive capacity of the nation's industrial base—which by the 1920s was particularly geared toward the production of consumer durables (items expected to last three years or more), and especially automobiles. As I note later, urban sprawl contributed significantly to creating the consumer-durables revolution in the United States. The centrality of urban sprawl to the U.S. economy was evident in the 1970s with the oil shocks. In spite of the severe economic and geopolitical vulnerability created by the oil dependency of urban sprawl, the U.S. government did not seek to curb this sprawl. Instead it responded militarily and diplomatically to ensure the ample flow of petroleum, as I demonstrate in the final section of the chapter.

Raw Materials and Marx's Theory of Value

At the heart Marx's political and economic analysis of capitalism is his claim that exchange value is solely derived from socially necessary labor. From this follows his key claim that capitalists attain their dominant economic, political, and social position by exploiting workers for their labor. While workers, through their labor, create all exchange value, workers only receive that part of the (exchange) value they create that is needed for their biological survival (necessary value). The capitalist retains the remainder, known as excess (or surplus) value.

Hence, according to Marxian political economy, raw materials have no intrinsic exchange value—again, because within capitalism all exchange value is produced by socially necessary human labor. As a result, the sale of natural resources does not contribute to wealth (i.e., value) creation within capitalism. Beyond the labor required to bring raw materials to market, any money/profit derived from the vending of raw materials is deemed a rent by Marx. In other words, money/profit gained through the selling of a raw material tends to be the result of strategic control or domination of supply. Money gained in such instances is a transfer of capital and not capital creation. It is a transfer of money (i.e., capital) from a capitalist and/or worker to the controller(s) of the raw material (including land).[9]

So-called free market environmentalists tacitly accept Marx's view that natural resources have no intrinsic exchange value. These environmental thinkers advocate the creation of rents for all natural resources. Free market environmentalists argue that the proper husbandry of natural resources can only arise through self-interest. As a result, because natural resources are provided free by nature (e.g., air and water), no one has any incentive to conserve or protect them (i.e., the tragedy of the commons)[10]—save for those individuals that can capture rent from their use. Therefore, if people could be assigned specific ownership over all natural resources, including the air and water, then they could charge a rent for their use. Owners would then have an incentive to protect and conserve all natural resources.[11]

Cyrus Bina explains that among classical political economists it is Marx's economic conceptions of natural resources and rent that provide the deepest insight into operation of the global petroleum market.[12] Neoclassical economists register prohibitively high rents as creating market disequilibrium (e.g., the oil crises of the 1970s), and count rents themselves as causing partial market disequilibrium (e.g., land in highly congested areas). Therefore, these thinkers do not analytically explain rents, but solely try to document their impact. Prior to Marx, it was David Ricardo who posited an argument to explain and measure rent in the market economy.[13] Like Marx, Ricardo held that labor was the source of wealth within capitalism. Also like Marx, he held that landed interests extracted rents from industrial capitalists and workers through agricultural commodities. In a key divergence from Marx, however, Ricardo reasoned that the land with the lowest productivity extracted zero rent and served as a baseline, whereas it was the more productive lands that extracted rent at an increasing scale. Bina centers his analysis of petroleum rent on the Persian Gulf region.[14]

The U.S. Petroleum Industry

The history of the U.S. petroleum industry demonstrates the validity and utility of Marx's economic conception of natural resources. From the late nineteenth to the middle of the twentieth century, the United States was the world's leading producer of oil. The U.S. firm Standard Oil was the first to globalize the trade in petroleum products—in particular kerosene, used for indoor lighting. This trade was initially based on Pennsylvania oil production. Later oil finds in the states of California,

Oklahoma, Texas, Indiana, and Louisiana established the United States as the world's prime source of oil through to the 1950s.[15]

Overproduction had been an issue for the U.S. oil industry almost from its inception. Entry into the U.S. oil extraction business in the nineteenth century was relatively easy. Standard Oil came to dominate the oil industry in this era not through extraction, but through petroleum refining. Because U.S. oil fields were mostly newly tapped, the natural pressure in the fields was high and the oil literally came out on its own. Therefore, the cost for producers to extract oil was basically nothing. Standard would buy inexpensive crude from producers and charge a premium price on refined petroleum products.[16] Overproduction issues were exacerbated because of new oil finds. Particularly devastating for the oil industry was a major discovery in Texas in the early 1930s.[17]

The industry's overproduction questions were partially the result of the courts' refusal to establish ownership over petroleum reservoirs—hence it was impossible to establish strategic control over them and use that control to manipulate supply. Anyone who could tap into a reservoir could extract as much oil from it as they desired, as long as they held legal title to the land on which they had established their well.[18] This served as an anticonservation measure, because no one knew if someone else was extracting petroleum from the same reservoir they were working. Therefore, the logical course of action was to extract all the oil from a well as quickly as possible, since someone else may draw the same oil from a different well.

During the late nineteenth century the U.S. timber industry found itself in a similar dilemma as U.S. oil producers. With the cost of trees at zero, and railroads reaching throughout the heavily forested Pacific Northwest, the timber market became badly glutted and the price of timber was perennially low. Unlike the case of oil production, the federal government during this period did intervene in an effort to push up timber prices. It did so through the U.S. Forest Service. On its creation in 1905, the service was given jurisdiction over the national forests, and these forests were greatly expanded. The national forests are generally of lower quality than privately held forestlands. Nevertheless, small timber operators would draw significant amounts of trees from them. Once these public forests came under the control of the Forest Service, access to the national forests was severely constricted.[19] It was not until the housing boom of the post–World War II era, however, that the price of timber attained a high level.[20]

Urban Sprawl and the U.S. Economy

The World War II U.S. housing boom came in the form of urban sprawl. The prices of petroleum and timber were low enough to allow for the sprawled quality of the U.S. housing boom because no one owner, or set of owners, of these abundant resources exercised dominant or strategic control over them (the Standard Oil trust was broken up in the 1910s). The prices of petroleum and timber were also sufficiently low because of the limited amount of labor needed to bring them to market.

The sprawled nature of the postwar housing boom is politically, economically, and, as it turns out, environmentally significant. This is because sprawl increases demand for automobiles. Indeed, urban sprawl makes an automobile a necessity.[21] Also, because urban sprawl tends to result in relatively large living abodes, sprawl also increases demand for such consumer durables as furniture and appliances. The higher energy demands of these larger abodes resulting from increased heating and cooling, as well as extra appliances and lighting, could be met because of the ample amounts of coal, natural gas, and uranium in the United States, which had a relatively decentralized ownership pattern.[22]

The techniques of urban sprawl were initially developed in the United States by landowners and developers who sought to bring utility to their landholdings on the urban periphery.[23] Early sprawl efforts begin in earnest during the late nineteenth century with the electric streetcar or trolley.[24]

With the advent of the automobile, its declining expense, and the growing public confidence in it during the 1910s and 1920s,[25] land developers began to develop land away from trolley lines in the urban periphery. This trend was most pronounced in Los Angeles.[26] Mark Foster,[27] a historian of U.S. urban transit, explains that the automobile "exerted a dramatic effect on the remote areas [of Los Angeles] which were not so well served by the trolleys." He goes on to explain that the Los Angeles "real estate boom of the 1920s witnessed the promotion of thousands of lots, many located miles from the nearest trolley lines."[28]

By the end of the 1920s, the Los Angeles area had become the U.S. region most adapted to the automobile, whereby "residents of Los Angeles purchased more automobiles per capita than did residents of any other city in the country." During this period "there were two automobiles for every five residents in Los Angeles, compared to one for every four residents in Detroit, the next most 'automobile oriented' American city."[29]

The Federal Government and Urban Sprawl

Beginning in the 1930s the federal government promoted urban sprawl as a means to resuscitate the U.S. economy. During this period the federal government initiated a program to underwrite home mortgages. It did so through the Federal Housing Authority (FHA). The FHA's legislative authority is found in the National Housing Act of 1934. The committee that composed this act was headed by Marriner Eccles, a wealthy Utah businessperson, who was an official in the Department of Treasury, and also included Albert Deane, assistant to the president and chairman of General Motors—Alfred Sloan.[30] Eccles's committee was actually a sub-committee of the President's Emergency Committee on Housing, which included W. Averell Harriman, who was asked to participate on this issue because of his "national standing as a businessman."[31] As historian Sydney Hyman explains, "When the terms of the new housing program were finally agreed to, [Harriman] was expected to 'sell' the program to . . . the business community at large."[32] Also on the President's Emergency Committee on Housing was John Fahey, chairman of the Federal Home Loan Bank Board.[33]

Eccles's biographer outlines the thinking underlying the formulation of the National Housing Act: "A program of new home construction, launched on an adequate scale, would not only gradually provide employment for building trade workers" but more importantly "accelerate the forward movement of the economy as a whole." It was anticipated that "its benefits would extend to everyone, from the manufacturers of lace curtains to the manufacturers of lumber, bricks, furniture, cement and electrical appliances."[34] Therefore, the purpose of the legislation that authorized the FHA was seemingly to spur consumption, including that of consumer durables. Urban sprawl would presumably help accomplish this goal, since by the 1920s, suburban developers had already demonstrated a predilection for building large, relatively expensive homes on undeveloped tracts of land far from trolley lines.[35]

Consumer Durables and Urban Sprawl in the United States

By the 1920s the United States was leading a consumer-durables revolution. As noted, consumer durables are goods expected to last at least three years. Economic historian Peter Fearon says of the other leading industrial power in the 1920s, Great Britain, that its "economy was retarded by the weight of the old staple industries such as cotton textiles, coal, shipbuilding and iron and steel."[36] He explains that this is "in contrast to the striking advance of the consumer-durables sector in

America."[37] Thus, the U.S. economy excelled in the production of such commodities as household appliances.[38]

The most prominent feature of the consumer-durables-geared U.S. industrial base was automobile production. In 1920 U.S. automobile firms produced 1.9 million automobiles, and in 1929 4.4 million. This represented 85 percent of total global automotive production. Fearon explains that "the influence of the automobile [on the U.S. economy] was pervasive." For example, "it provided one of the chief markets for the steel industry and for the manufacturers of glass and tires."[39] During much of the 1920s "nearly 17 percent of the total value of fully and semi-manufactured goods was accounted for by automotive products."[40] Statistics like these prompt economic historian Elliot Rosen to regard the automotive industry as the "nation's principal industry" by the 1920s.[41] Another economic historian, Richard B. Du Boff, notes that "during the 1920s, the [automotive] industry became the nation's leader in manufacturing."[42]

While the productive capacity of automotive manufacturers greatly expanded throughout the 1920s and automobile production had significant implications for overall industrial activity in the United States, the demand for automobiles would greatly fluctuate. The overall trend in automotive production during the 1920s was upward, but market downturns caused significant production declines in 1921, 1924, and 1927.[43] Additionally, during earlier recessions automobile output "contracted severely."[44] During the Great Depression, among industrial producers "the collapse in the motor vehicle sector was especially pronounced." By the end of 1929 "the reduction in automobile output was the greatest in the entire manufacturing sector."[45]

On its creation, the FHA was placed under the stewardship of prominent officials from the real estate sector, and they used their authority to promote the horizontal growth of urban America. Created in 1934, "FHA's staff was recruited almost entirely from the private sector. Many were corporate executives from a variety of different fields, but real estate and financial backgrounds predominated."[46] Jeffrey Hornstein, a historian of the U.S. real estate industry, notes that the industry generally "welcomed the FHA ... both because it promised greatly enhanced general demand for housing and because the agency was run largely by Realtors and their allies in the banking world."[47]

As a way to encourage housing sales, the FHA underwrote home purchases. It would guarantee 80 percent of home mortgages for qualified homes and buyers for a twenty-year term. (Later, this guarantee was modified to 90 percent and twenty-five years.) Up to this time, standard mortgages covered about 50 percent of the home purchase price and had a three-year term.[48]

This program gave the FHA the ability to influence the types of homes purchased and, subsequently, housing development patterns. Mare Weiss, in his history of the U.S. real estate industry, notes that "because FHA could refuse to insure mortgages on properties due to their location in neighborhoods that were too poorly planned or unprotected and therefore too 'high-risk', it definitely behooved most reputable subdividers to conform to FHA standards."[49] With this power, the FHA promoted the building of large-scale housing developments in outlying areas. Weiss explains that the Federal Housing "Administration's clear preference . . . was to use conditional commitments [for loan guarantees] specifically to encourage large-scale producers of complete new residential subdivisions, or 'neighborhood units'." Thus, the FHA, through its loan program, encouraged and subsidized "privately controlled and coordinated development of whole residential communities of predominately single-family housing on the urban periphery."[50]

Kenneth Jackson, in his important history on the suburbanization of urban development in the United States, concurs with Weiss's assessment of FHA bias toward new housing stock in outlying areas.[51] Jackson writes that "in practice, FHA insurance went to new residential developments on the edges of metropolitan areas, to the neglect of core cities."[52] As a result, he notes that between 1942 and 1968 the "FHA had a vast influence on the suburbanization of the United States."[53]

The Consumer-Durables Revolution

In her historical analysis of U.S. consumption patterns, economic historian Martha L. Olney finds that "between 1919 and 1928, [U.S. households] spent annually an average of $267 each on durable goods—$172 for major durables (now mostly automobiles and parts rather than furniture) and only $96 for minor durables (still mostly china and tableware, house furnishings, and jewelry and watches)."[54] After a number of decades of urban horizontal growth,[55] "By 1979–86, households annually spent an average of $3,271 each for durable goods, with $2,230 for major durables (still predominantly automobiles and parts) and $1,041 for minor durable goods (now house furnishings, miscellaneous other durable goods, and jewelry and watches)."[56] Conveyed in constant dollars, households spent an average of $955 on consumer durables between 1919 and 1928, and $3,353 between 1979 and 1986.[57] Olney adds that "strong growth purchases of automobiles and parts remain evident: average annual purchases for 1919–28 were four times greater than the average for 1909–18, and growth continued through the post–World War II years." Additionally, "purchases of household appliances and the 'entertainment complexes'—radios, televisions, pianos, and other musical instruments—showed a similar pattern."[58]

Utilizing statistical analysis, Olney demonstrates that the dramatic increases in the consumption of durable goods exceeded overall increases in income during the pre–Depression Era and the post–World War II period.[59] For this reason, she contends that the 1920s mark the beginning of the consumer-durables revolution in the United States. She attributes the surges in the consumption of consumer durables to two factors: advertising and the availability of consumer credit. She acknowledges, however, that advertising,[60] and especially consumer credit,[61] were not as widespread during the 1920s as they were after World War II.[62] What was evident during both of these periods was an increasing trend of urban sprawl, expanding the demand for consumer durables.

Today, U.S. urban sprawl has international economic ramifications. The United States is the world's largest consumer.[63] Importantly, European, Japanese, and South Korean automakers count heavily on access to the huge U.S. automobile market to attain profitability.[64]

Post–World War II Urban Sprawl and U.S. Oil Policy

In 1973 the Persian Gulf region of the Middle East took on particular importance for the Western allies. What came into relief in 1973 is that the region contained the key supplies of petroleum for the Western world. The petroleum-bearing countries of the region are Iran, Iraq, Kuwait, Saudi Arabia, United Arab Emirates, and Qatar, with Iran, Iraq, Kuwait, and Saudi Arabia being the primary producing countries for the world's oil market. These latter nations today possess the majority of the world's known petroleum reserves—Saudi Arabia alone is estimated to hold 25 percent of the world's proven reserves of petroleum.[65]

The Persian Gulf's strategic importance is in significant part the result of U.S. oil policies. This is particularly apparent on the demand side. As U.S. cities became more and more sprawled,[66] and as a result more automobile dependent,[67] U.S. oil consumption steadily climbed.[68] Between 1946 and 1953, for instance, U.S. gasoline usage went from 30 billion gallons annually to 49 billion, amounting to a yearly growth rate of slightly over 7.2 percent. In 1958 U.S. gasoline consumption exceeded 59 billion gallons.[69]

U.S. consumption had a detrimental effect on its petroleum production. This was important because the United States was historically capable of reducing world petroleum prices through increased production. By 1970, however, U.S. oil production had peaked, and it was no longer capable of regulating world prices.[70] When Saudi Arabia imposed a selective embargo on countries favorable to Israel in 1973, the United States was importing close to 40 percent of its oil needs, and it could not respond to the shortfall created by the embargo with domestic production.[71]

The Oil Shocks of the 1970s

Therefore, leading up to the oil shocks of the 1970s, U.S. oil reserves were depleted for two key reasons: an underdeveloped legal regime, and high levels of domestic consumption. What is theoretically and historically significant, however, is the response of the U.S. government when the dependency and vulnerability of the U.S. economy on foreign sources of petroleum came into stark relief in 1973. No effort was put forward by the U.S. government to roll back or limit urban sprawl and the automobile dependence that it spawned.[72]

The United States responded diplomatically and militarily to its apparent dependency. U.S. policymakers used the country's superior political and military position to ensure that Persian Gulf oil remained in the U.S. sphere of influence, and that the region's petroleum flowed sufficiently. Until 1979, the United States amply supplied the Iranian government with military equipment and training to safeguard the petroleum reserves of the region against any Soviet aggression. After its client regime in Iran collapsed (which brought on a second oil crisis), the United States sought to directly build up its military capabilities in the region, culminating with a direct military presence after the first Persian Gulf War in 1991.[73]

This emphasis on the supply side to deal with the U.S. energy problems of the 1970s is reflected in two reports put out by the Twentieth Century Fund (now the Century Fund). This organization is a foundation, which in the 1950s and 1960s sponsored studies on the natural resource needs of the expanding U.S. economy.[74] The Twentieth Century Fund created two policy groups in the early 1970s that put forward proposals to deal with the U.S. petroleum situation. One task force, convened in 1973, was titled "The Twentieth Century Fund Task Force on United States Energy Policy." On this task force was a director and senior vice president of Exxon; a vice chairman of the board of the American Electric

Power Company; Walter J. Levy (a consultant to most major oil firms);[75] a vice chairman of the board of Texas Commerce Bancshares (a major Texas bank);[76] and the chairman of the board of Carbomin International Corporation (an international mining firm). The other task force, formed in 1974, was known as "The Twentieth Century Fund Task Force on the International Oil Crisis." Walter J. Levy, the executives from Carbomin, and those from Texas Commerce Bancshares also served on this task force. Others on this Twentieth Century Fund task force included the chairman of the board of Atlantic Richfield (an oil firm), a managing director of Dillon, Read & Co. (a leading New York investment management firm), the chairman of the board of the Louis Dreyfus Corporation (an investment management firm), the chairman and president of the First National Bank of Chicago, and a consultant to Wells Fargo Bank (a major California bank). Also on these task forces were academics (mostly economists) from Princeton, Harvard, MIT, and the University of Virginia, as well as the presidents of Resources for the Future and the Carnegie Institution, both of which are economic elite–led research institutes. (The president of Resources for the Future served on both task forces; the Carnegie Institution president was a member only of the energy policy group).[77]

In the wake of the 1973 oil shortage and the efforts of the Organization of Petroleum Exporting Countries (OPEC) to maintain high oil prices, both of the Twentieth Century Fund's task forces advised that the United States should strive to develop sources of oil and energy outside of the OPEC countries. This would reduce the strategic positioning of OPEC countries over petroleum and petroleum prices. OPEC includes all the Persian Gulf oil producers, plus Algeria, Angola, Libya, Nigeria, Venezuela, and Indonesia. The Twentieth Century Fund's task force on the international oil crisis advised that "the best remedy for the problems caused by the increased price of oil [brought about by OPEC members] would be, simply, to lower the price" of petroleum. "The Task Force believes that this remedy should be sought through reliance on market forces."[78] The task force goes on to explain in its report that "*the most effective means of exerting market pressure will be to accelerate exploration for crude and develop producing capacity*" from areas outside of OPEC.[79] The task force on U.S. energy policy averred "*that it is essential that the nation take firm and forceful action to implement a comprehensive near-term energy program designed to assure greater availability of domestic supplies of oil and other sources of energy.*"[80] Therefore, in light of U.S. oil dependency on OPEC countries, the key recommendations put forward by these policy groups—made up in large part of economic elites—was to expand the supply of available petroleum free from OPEC control, thereby weakening the strategic control of oil held by OPEC nations. This would lessen the rent they could charge for petroleum.

Both these groups, in their reports, called for greater energy efficiency, or what they labeled in their reports as "conservation." The difficulty is that increased energy efficiency does not necessarily reduce overall consumption levels. The energy policy group, in a section of its report titled "Measures to Promote Conservation", "*endorse[d] the use of special incentives to encourage further investment in energy-saving capital goods and consumer durables because conserving energy is as important as increasing the supply.*"[81] In its report, it specifically suggested the use of a "luxury" tax to discourage the purchase of large, less efficient automobiles. Moreover, the implementation of "excise taxes levied annually and collected with state registration fees also might serve to encourage quicker scrapping of cars that consume above-average amounts of gasoline."[82] Finally, the "Task Force favor[ed] the continuation of such energy-conserving measures as reasonable speed limits on highways."[83] The task force on the international oil crisis did not set out specific conservation proposals. Instead, it deferred to the energy policy task force on this.[84]

Increased energy efficiency can lead to overall lower levels of petroleum consumption. Energy savings from increased efficiency, however, can be offset by increased economic growth. This is especially the case within sprawled urban regions, where greater levels of economic activity can lead to a larger workforce driving to and from work, and increased demand for spacious homes on the urban periphery. So whereas automobiles may become more fuel efficient, in the context of diffusely organized cities more automobiles and longer driving distances can lead to greater overall gasoline/oil consumption—in spite of gains made in fuel efficiency. This is precisely what has transpired in the United States. The current U.S. automobile fleet is more efficient than the U.S. automotive fleet of the early 1970s.[85] Because of a substantially enlarged automobile population and ever-increasing amounts of driving, however, gasoline/diesel consumption in the United States today substantially exceeds that of the 1970s. According to energy economist Ian Rutledge, in 1970 automobile driving in the United States consumed 7.1 million barrels per day of petroleum, whereas by 2001 that figure increased to 10.1 million.[86] Today, according to U.S. government agencies, automobile driving in the United States consumes over 10 percent of total global oil production.[87] In large part because of the steady growth of gasoline/diesel consumption in the United States,[88] its economy consumes 25 percent of the world's total petroleum.[89] This is especially glaring, because in the aftermath of the spike in oil prices in the 1970s, U.S. factories and utilities shifted from petroleum-based fuels to other sources of energy (mostly coal, natural gas, and nuclear power).[90] It is telling that neither of the Twentieth Century Fund's task forces counseled less driving or mass transportation as conservation measures to counter OPEC price strategies. Such a recommendation would have raised urban sprawl and the automobile dependence that it creates as political issues.

Conclusion

Treating natural resources as having zero exchange value provides substantial insight into why U.S. urban zones are the most sprawled in the world.[91] Because of this sprawl the United States is the highest global absolute and per capita emitter of the primary greenhouse gas—carbon dioxide.[92] Beginning with the advent of the trolley in the late nineteenth century, land developers in the United States started developing significant amounts of land on the urban periphery as low-density residential areas. A specific advantage for developers of low-density housing developments is that they are selling—for a significant price—a commodity that at least initially cost little to nothing: open land. Low-density housing tracts have a great deal of minimally developed acreage in the form of yards. In the 1920s, the automobile came fully "online" as a mass consumption item, and land developers accelerated the development of land distant from city centers—because suburban land away from trolley lines could now be commercially developed.

Such land development patterns in the United States during the early automobile age could be pursued because of the abundance of oil supplies in the United States, their easy extraction, and their diffuse ownership. As a result, gasoline prices were persistently low, and the growing number of suburban residents could afford their automobile dependency.

Hence, the development of residential neighborhoods on the urban periphery expanded the demand for automobiles. By the 1920s the automobile was already a key product of the U.S. industrial base. The relationship between suburban development and automobile consumption can be observed in the case of Los Angeles in the 1920s. During this period, Los Angeles land developers were national leaders in building tracts of housing on the urban outskirts. Not surprisingly, Los Angeles during this time had a higher per capita automobile ownership than any other major urban area in the United States.

In the midst of the Great Depression, in the 1930s, the federal government launched a housing program in order to resuscitate the U.S. economy. Marriner Eccles's biographer notes that a specific purpose of the federal housing program was to increase the demand for consumer durables—because new home purchases would increase the demand for household items. Eccles, a Treasury official at the time, was on the President's Emergency Committee on Housing. This committee formulated the National Housing Act of 1934. From this law, the Federal Housing Authority or FHA was created. The FHA subsidized home purchases by guaranteeing mortgages. The bias of the FHA was to guarantee single-family home purchases on the urban periphery. Not only would spacious homes in suburban areas spur the consumption of household items, but such homes would necessitate automobile ownership. Olney outlines how consumer-durables consumption in the post–World War II period, including that of automobiles, surged.[93] It was during this period that FHA policies were aggressively pushing urban regions in a horizontal direction.

This history of urban sprawl policies leading to increased consumer-durables consumption is consistent with Marx's contention that within capitalism, economic demand is shaped to maximize the realization of profit. In contrast, neoclassical economic philosophers tend to take economic demand as a given, and instead focus on how producers respond to such demand.[94]

The FHA's pro–urban sprawl policies would have the additional effect of increasing demand for petroleum, the price of which was severely depressed in the 1930s—due at least in part to massive new domestic finds and a poorly developed ownership regime. By the early 1970s, however, U.S. domestic supplies of oil were exhausted. As a result, OPEC came to exercise strategic control over the world's oil supplies. The response of the U.S. government was to implement policies that would seek to minimize the rent that OPEC producers could charge for their petroleum by using its superior military and political position to ensure that ample supplies of oil flowed onto the international market.

As outlined throughout this chapter, the pro–urban sprawl policies of the U.S. government have been consistent with the interests and, seemingly, the policy preferences of economic elites and producer groups.

Economic elites were on the President's Emergency Committee on Housing. (Most noteworthy was the presence on the committee of Alfred Sloan—president and chairperson of General Motors at the time—through his assistant.) The FHA was placed under the authority of officials from the real estate industry. Finally, the U.S. foreign oil policies in the aftermath of the oil crisis of 1973 are consistent with the policy advice put forward by the Twentieth Century Fund's task forces—both of which were composed largely of economic elites. The focus of this advice was to deal with the oil crisis predominantly through supply-side policies, and not by reducing the automobile dependency brought on by urban sprawl. The results of these policies have been more petroleum consumed by U.S. drivers than before the oil crisis of 1973, and the unabated emission of anthropogenic climate change gasses (especially carbon dioxide) by the U.S. economy.

Notes

1. Jeffrey R. Kenworthy and Felix B. Laube, with Peter Newman, Paul Barter, Tamim Raad, Chamlong Poboon, and Benedicto Guia Jr., *An International Sourcebook of Automobile Dependence in Cities 1960–1990* (Boulder: University Press of Colorado, 1999). Two key indicators of urban sprawl and automobile dependency are per capita automobile ownership and automobile usage. In Kenworthy's and Laube's forty-six-international-city study they found that the U.S. cities in their study together had the highest total figures on both counts. In the U.S. cities in these authors' study, there were 604 automobiles per 1,000 people. For the other cities the number of automobiles per 1,000 individuals were as follows: Australian cities, 491; Canadian, 524; European, 392; wealthy Asian, 123; and developing Asian, 102. Each automobile in the U.S. cities studied was driven on average for 11,155 kilometers. In Australia this average was 6,571; Canada, 6,551; Europe, 4,519; wealthy Asian cities, 1,487; and developing Asian cities, 1,848 (529). The ratio of the average use of each automobile in the U.S. cities compared to the others was: Australia, 1.70; Canada, 1.70; Europe, 2.47; wealthy Asian cities, 7.50; and developing Asian cities, 6.04 (Kenworthy and Laube et al., *An International Sourcebook of Automobile Dependence in Cities*, 530).

2. For example, Dolores Hayden, *Building Suburbia* (New York: Pantheon, 2003); Jennifer Wolch, Manuel Pastor Jr., and Peter Drier, eds., *Up against the Sprawl* (Minneapolis: University of Minnesota Press, 2004); George A. Gonzalez, *The Politics of Air Pollution* (Albany: State University of New York Press, 2005); George A. Gonzalez, "Urban Sprawl, Global Warming, and the Limits of Ecological Modernization," *Environmental Politics* 14, no. 3 (2005): 344–362. George A. Gonzalez, *Urban Sprawl, Global Warming, and the Empire of Capital* (Albany: State University of New York Press, forthcoming); Robert D. Bullard, ed., *Growing Smarter: Achieving Livable Communities, Environmental Justice, and Regional Equity* (Cambridge, MA: MIT Press, 2007).

3. David Goodstein, *Out of Gas: The End of the Age of Oil* (New York: Norton, 2004); Paul Roberts, *The End of Oil* (New York: Houghton Mifflin, 2004).

4. Howard L. Parsons, "Introduction," in Howard L. Parsons, ed. and comp., *Marx and Engels on Ecology* (Westport, CT: Greenwood Press, 1977); Paul Burkett, *Marx and Nature* (New York: St. Martin's Press, 1999); Jonathan Hughes, *Ecology and Historical Materialism* (New York: Cambridge University Press, 2000).

5. In the *Grundrisse* Marx lists three ways in which the "production of surplus value [within capitalism], based on the increase and development of the productive forces, requires the production of new consumption; requires that the consuming circle within circulation expands as did the productive circle previously." This is accomplished "firstly, [through] quantitative expansion of existing consumption; secondly: creation of new needs by propagating existing ones in a wide circle; thirdly: production of new needs and discovery and creation of new use values" (Karl Marx, *Grundrisse* (New York: Vintage, 1973), 408).

6. Robert Paul Wolff, *Understanding Marx: A Reconstruction and Critique of Capital* (Princeton, NJ: Princeton University Press, 1984); Duncan K. Foley, *Understanding Capital: Marx's Economic Theory* (Cambridge, MA: Harvard University Press, 1986).

7. Ralph Miliband, *The State in Capitalist Society* (New York: Basic Books, 1969); Mancur Olson, *The Logic of Collective Action* (Cambridge, MA: Harvard University Press, 1971); John F. Manley, "Neo-Pluralism," *American Political Science Review* 77, no. 2 (1983): 368–383; Clyde W. Barrow, *Critical Theories of the State* (Madison: University of Wisconsin Press, 1993), chap. 1; George A. Gonzalez, "Ideas and State Capacity, or Business Dominance? A Historical Analysis of Grazing on the Public Grasslands," *Studies in American Political Development* 15 (2001): 234–244; George A. Gonzalez, *Corporate Power and the Environment* (Lanham, MD: Rowman & Littlefield, 2001); George A. Gonzalez, "The Comprehensive Everglades Restoration Plan: Economic or Environmental Sustainability?", *Polity* 37, no. 4 (2005): 466–490; Andrew S. McFarland, *Neopluralism* (Lawrence: University Press of Kansas, 2004); G. William Domhoff, *Who Rules America?* (New York: McGraw-Hill, 2005); Paul Wetherly, Clyde W. Barrow, and Peter Burnham, eds., *Class, Power and the State in Capitalist Society: Essays on Ralph Miliband* (New York: Palgrave MacMillan, 2008). The U.S. economic elite is composed of decision makers within large corporations and of other persons of substantial wealth. These actors are integrated into a cohesive elite or class through social clubs, interlocking directorates of both private and public organizations, policy discussion groups, and intermarriage. See Michael Useem, *The Inner Circle: Large Corporations and the Rise of Business Political Activity in the U.S. and U.K.* (Oxford: Oxford University Press, 1984); Clyde W. Barrow, *Critical Theories of the State*, chap. 1; G. William Domhoff, *Who Rules America?*. Altogether, the economic elite compose roughly 0.5 to 1 percent of the total U.S. population. This elite is a dominant factor in the development of public policy because it possesses the most important political resources in the United States—wealth and income. Their superior amounts of wealth and income allow members of the economic elite superior access to cam-

paign finance, organization, publicity, lobbying, and scientific and legal expertise. All of these "political tools" are used to shape public policies (Barrow, *Critical Theories of the State*, 16–17).

Given their substantial wealth, income, and high level of organization, producer groups are formidable political actors on their own. See Olson, *The Logic of Collective Action*; Charles E. Lindblom, *Politics and Markets: The World's Political-Economic Systems* (New York: Basic Books, 1977); Barrow, *Critical Theories of the State*, chap. 1. Due to shared ideological, economic, and political interests, however, members of the economic elite are normally able to draw in as political allies members of broad-based producer groups, and vice versa (Domhoff, *Who Rules America?*).

8. In contrast, the other leading industrial capitalist regions of the world (e.g., Western/Central Europe and Japan) have historically had little domestic production of petroleum.

9. Burkett, *Marx and Nature*, chap. 6; Paul Burkett, "Nature's 'Free Gifts' and the Ecological Significance of Value," *Capital & Class* 68 (1999): 89–110; Paul Burkett, *Marxism and Ecological Economics: Toward a Red and Green Political Economy* (Boston: Brill, 2006).

10. Garrett Hardin, "The Tragedy of the Commons," *Science* 162 (1968): 1243–1248.

11. Terry L. Anderson and Donald R. Leal, *Free Market Environmentalism* (New York: Palgrave, 2001); John Dryzek, *The Politics of the Earth*, 2nd ed. (New York: Oxford University Press, 2005), chap. 6.

12. Cyrus Bina, "Some Controversies in the Development of Rent Theory: The Nature of Oil Rent," *Capital & Class* 39 (1989): 82–112.

13. David Ricardo, *On the Principles of Political Economy and Taxation* (Washington, DC: J. B. Bell, 1830).

14. Cyrus Bina, *The Economics of the Oil Crisis* (New York: St. Martin's Press, 1985).

15. Daniel Yergin, *The Prize: The Epic Quest for Oil, Money, and Power* (New York: Simon & Schuster, 1991).

16. Yergin, *The Prize*, chap. 2.

17. David Davis, *Energy Politics* (New York: St. Martin's Press, 1993), chap. 3; Steve Isser, *The Economics and Politics of the United States Oil Industry, 1920–1990* (New York: Garland, 1996); Diana Davids Olien, and Roger M. Olien, *Oil in Texas: The Gusher Age, 1895–1945* (Austin: University of Texas Press, 2002).

18. Stephen L. McDonald, *Petroleum Conservation in the United States: An Economic Analysis* (Baltimore: Johns Hopkins University Press, 1971); Edward Miller, "Some Implications of Land Ownership Patterns for Petroleum Policy," *Land Economics* 49, no. 4 (1973): 414–423.

19. William G. Robbins, *Lumberjacks and Legislators: Political Economy of the U.S. Lumber Industry, 1890–1941* (College Station: Texas A&M University Press, 1982); George A. Gonzalez, "The Conservation Policy Network, 1890–

1910: The Development and Implementation of 'Practical' Forestry." *Polity* 31, no. 2 (1998): 269–299; Gonzalez, *Corporate Power and the Environment*, chap. 2.

20. Randal O'Toole, *Reforming the Forest Service* (Washington, DC: Island Press, 1988); Paul W. Hirt, *A Conspiracy of Optimism: Management of the National Forests Since World War Two* (Lincoln: University of Nebraska Press, 1994).

21. Peter Newman and Jeffrey Kenworthy, *Sustainability and Cities: Overcoming Automobile Dependence* (Washington, DC: Islands Press, 1999).

22. Richard H. Vietor, *Environmental Politics & the Coal Coalition* (College-Station: Texas A&M University Press, 1980); David Davis, *Energy Politics*; Harvey Blatt, *America's Environmental Report Card* (Cambridge, MA: MIT Press, 2005), chap. 5. Energy expert Paul Roberts explains that contemporary U.S. households are "at least twice as energy-intensive as European and Japanese households." Paul Roberts, *The End of Oil*, 152; also see Martin Fackler, "The Land of Rising Conservation: Japan Offers a Lesson in Using Technology to Reduce Energy Consumption" *New York Times*, Jan. 5 2007, *p.* C1.

23. Marc Weiss, *The Rise of the Community Builders: The American Real Estate Industry and Urban Land Planning* (New York: Columbia University Press, 1987); Jeffrey M. Hornstein, *A Nation of Realtors: A Cultural History of the Twentieth-Century American Middle Class* (Durham, NC: Duke University Press, 2005).

24. Mark S. Foster, "The Model-T, the Hard Sell, and Los Angeles's Urban Growth: The Decentralization of Los Angeles during the 1920s," *Pacific Historical Review* 44 (1975): 459–484; Gonzalez, *The Politics of Air Pollution*, chap. 4.

25. James Flink, *The Car Culture* (Cambridge, MA: MIT Press, 1975); James Flink, *The Automobile Age* (Cambridge, MA: MIT Press, 1990).

26. Weiss, *The Rise of the Community Builders*; Greg Hise, *Magnetic Los Angeles: Planning the Twentieth-Century Metropolis* (Baltimore: Johns Hopkins University Press, 1997).

27. Mark S. Foster, *From Streetcar to Superhighway: American City Planners and Urban Transportation, 1900–1940* (Philadelphia: Temple University Press, 1981).

28. Foster, "The Model-T, the Hard Sell, and Los Angeles's Urban Growth," 477.

29. Foster, "The Model-T, the Hard Sell, and Los Angeles's Urban Growth," 483.

30. Sidney Hyman, *Marringer S. Eccles* (Stanford, CA: Stanford University Graduate School of Business, 1976), 144.

31. Hyman, *Marringer S. Eccles*, 142; also see Rudy Abramson, *Spanning the Century: The Life of W. Averell Harriman, 1891–1986* (New York: Morrow, 1992).

32. Hyman, *Marringer S. Eccles*, 142.

33. Hyman, *Marringer S. Eccles*, 142. This board headed up the Federal Home Loan Bank System, created in 1932. It was made up of eleven regionally based home loan banks that served as a central credit agency similar to the Federal Reserve System. (Hyman, *Marringer S. Eccles*, 140).

34. Hyman, *Marringer S. Eccles*, 141. General Motors's presence on a government housing committee seeking the expansion of consumer durables consumption contravenes economic historian Elliot Rosen's claim that during the Great Depression "there is scant evidence, if any, that automobile producers, the nation's principal industry, sought government intervention." Rosen goes on to assert that "in the automobile industry" the market was permitted to operate unimpeded and unaided" (Elliot Rosen, *Roosevelt, the Great Depression, and the Economics of Recovery* (Charlottesville: University of Virginia Press, 2005), 118).

35. Robert Fishman, *Bourgeois Utopias: The Rise and Fall of Suburbia* (New York: Basic Books, 1987); Weiss, *The Rise of the Community Builders*; Robert M. Fogelson, *Bourgeois Nightmares: Suburbia, 1870–1930* (New Haven, CT: Yale University Press, 2005).

36. Peter Fearon, *War, Prosperity and Depression: The U.S. Economy 1917–45* (Lawrence: University Press of Kansas, 1987), 48.

37. Fearon, *War, Prosperity and Depression*, 48.

38. Robert D. Atkinson, *The Past and Future of America's Economy* (Northampton, MA: Edward Elgar, 2004); Alexander J. Field, "Technological Change and U.S. Productivity Growth in the Interwar Years," *Journal of Economic History* 66, no. 1 (2006): 203–236.

39. Fearon, *War, Prosperity and Depression*, 55.

40. Fearon, *War, Prosperity and Depression*, 58; also see Jean-Pierre Bardou, Jean-Jacques Chanaron, Patrick Fridenson, and James M. Laux, *The Automobile Revolution* (Chapel Hill: University of North Carolina Press, 1982); David J. St. Clair, *The Motorization of American Cities* (New York: Praeger, 1986); Atkinson, *The Past and Future of America's Economy*; Matthew Paterson, *Automobile Politics* (New York: Cambridge University Press, 2007).

41. Rosen, *Roosevelt, the Great Depression, and the Economics of Recovery*, 118.

42. Richard B. Du Boff, *Accumulation & Power: An Economic History of the United States* (Armonk, NY: M. E. Sharpe, 1989), 83.

43. Fearon, *War, Prosperity, and Depression*, 58; also see Robert Paul Thomas, *An Analysis of the Pattern of Growth of the Automobile Industry, 1895–1929* (New York: Arno, 1977); Bardou et al., *The Automobile Revolution*, chap. 6.

44. Fearon, *War, Prosperity, and Depression*, 58.

45. Fearon, *War, Prosperity, and Depression*, 91; also Atkinson, *The Past and Future of America's Economy*.

46. Weiss, *The Rise of the Community Builders*, 146.

47. Hornstein, *A Nation of Realtors*, 150.

48. Weiss, *The Rise of the Community Builders*, 146.

49. Weiss, *The Rise of the Community Builders*, 148.

50. Weiss, *The Rise of the Community Builders*, 147; also see Hornstein, *A Nation of Realtors*, 150–152.

51. Kenneth T. Jackson, *Crabgrass Frontier: The Suburbanization of the United States* (New York: Oxford University Press, 1985).

52. Jackson, *Crabgrass Frontier*, 206.

53. Jackson, *Crabgrass Frontier*, 209.

54. Martha L. Olney, *Buy Now, Pay Later: Advertising, Credit, and Consumer Durables in the 1920s* (Chapel Hill: University of North Carolina Press, 1991), 9.

55. Peter Muller, *Contemporary Suburban America* (Englewood Cliffs, NJ: Prentice-Hall, 1981); Robert A. Beauregard, *When America became Suburban* (Minneapolis: University of Minnesota Press, 2006).

56. Olney, *Buy Now, Pay Later*, 9.

57. Olney, *Buy Now, Pay Later*, 9.

58. Olney, *Buy Now, Pay Later*, 22.

59. Olney, *Buy Now, Pay Later*.

60. Also see Michael Dawson, *The Consumer Trap: Big Business Marketing in American Life* (Chicago: University of Illinois Press, 2003); Cynthia Lee Henthorn, *From Submarines to Suburbs: Selling a Better America, 1939–1959* (Columbus: Ohio State University Press, 2006).

61. Also see Lendol Calder, *Financing the American Dream: A Cultural History of Consumer Credit* (Princeton: Princeton University Press, 1999); Rosa-Maria Gelpi and François Julien-Labruyère, *The History of Consumer Credit* (New York: St. Martin's Press, 2000), chap. 8; Rowena Olegario, *A Culture of Credit: Embedding Trust and Transparency in American Business* (Cambridge, MA: Harvard University Press, 2006).

62. Most automobiles in the 1920s were purchased through credit. Nonetheless, as consumer credit historian Lendol Caldwell explains "throughout the 1920s 25 to 40 percent of Americans in any given year continued to buy cars for cash." (Calder, *Financing the American Dream*, 194).

The terms of retail automotive credit were rather stringent during the 1920s. Loan terms required one-third of the purchase price upon signing, and the amortization period on automobile loans was from 6 to 12 months. Moreover, economic historian Martha Olney reports that the "effective annual interest rate exceeded 30 percent" on automobiles purchased through credit in the 1920s. (Martha L. Olney, "Credit as a Production-Smoothing Device: The Case of Automobiles, 1913–1938." *Journal of Economic History* 49, no. 2 (1989): 381). In the 1930s automotive credit terms were substantially liberalized. A general liberalization occurred in the 1930s on terms of credit with regard to the purchase of consumer durables (Calder, *Financing the American Dream*, 275).

63. Norman Frumkin, *Tracking America's Economy* (Armonk, NY: M. E. Sharpe, 2004). U.S. consumers, excluding government and businesses, purchase close to 20 percent of the world's total economic output, while comprising about 4.5 percent of the world's population (Peter S. Goodman, "The U.S. Economy: Trying to Guess what Happens Next," *New York Times*, Nov. 25, 2007, sec. 4, p. 1).

64. John A. C. Conybeare, *Merging Traffic: The Consolidation of the International Automobile Industry* (Lanham, MD: Rowman & Littlefield, 2004); Helmut Becker, *High Noon in the Automotive Industry* (New York: Springer, 2006); Martin Fackler, "Toyota Expects Decline in Annual Profit," *New York Times*, May 9, 2008, p. C3. The Japanese automakers Honda and Toyota (the world's second largest automobile manufacturer), for instance, derive two-thirds of their overall profits from sales in the United States (Todd Zaun, "Honda Tries to Spruce Up a Stodgy Image," *New York Times*, March 19, 2005, p. C3; Martin Fackler, "Toyota's Profit Soars, Helped by U.S. Sales," *New York Times*, August 5, 2006, p. C4).

65. Blatt, *America's Environmental Report Card*, 100; Francisco Parra, *Oil Politics* (New York: I. B. Tauris, 2005); John S. Duffield, *Over a Barrel: The Costs of U.S. Foreign Oil Dependence* (Stanford, CA: Stanford University Press, 2008).

66. Muller, *Contemporary Suburban America*.

67. Foster, *From Streetcar to Superhighway*; Kenworthy and Laube et al., *An International Sourcebook of Automobile Dependence in Cities 1960–1990*.

68. George Philip, *The Political Economy of International Oil* (Edinburgh: Edinburgh University Press, 1994); Ian Rutledge, *Addicted to Oil: America's Relentless Drive for Energy Security* (New York: I. B. Tauris, 2005); Duffield, *Over a Barrel*, chap 2.

69. American Petroleum Institute, *Petroleum Facts and Figures* (New York: American Petroleum Institute, 1959), 246–247.

70. Kenneth S. Deffeyes, *Hubbert's Peak: The Impending World Oil Shortage* (Princeton, NJ: Princeton University Press, 2001). U.S. petroleum production peaked at just under 10 million barrels a day in 1970, and in spite of recently increased drilling activity is down to 5.1 million barrels per day of production (Clifford Krauss, 2007 Nov. 2. "Tapped Out, but Hopeful: A Break in Texas's Oil Decline," *New York Times*, Nov. 2, 2007, p. C1).

71. John M. Blair, *The Control of Oil* (New York: Pantheon, 1976); Rutledge, *Addicted to Oil*.

72. Jon Van Til, *Living with Energy Shortfall* (Boulder, CO: Westview Press, 1982).

73. Steve A. Yetiv, *Crude Awakenings: Global Oil Security and American Foreign Policy* (Ithaca, NY: Cornell University Press, 2004); Steve A. Yetiv, *The Absence of Grand Strategy: The United States in the Persian Gulf, 1972–2005* (Baltimore: Johns Hopkins University Press, 2008); Rachel Bronson, *Thicker*

Than Oil: America's Uneasy Partnership with Saudi Arabia (New York: Oxford University Press, 2006); Duffield, *Over a Barrel.*

74. For example, Frederic Dewhurst and the Twentieth Century Fund, *America's Needs and Resources* (New York: Twentieth Century Fund, 1955); Thomas Reynolds Carskadon and George Henry Soule, *USA in New Dimensions: The Measure and Promise of America's Resources, A Twentieth Century Fund Survey* (New York: Macmillan, 1957); Arnold B. Barach and the Twentieth Century Fund, *USA and Its Economic Future* (New York: Macmillan, 1964).

75. In a 1969 profile of Walter J. Levy, titled "As Oil Consultant, He's without Like or Equal," the *New York Times* noted that "he is readily acknowledged as the 'dean of oil consultants' even by competitors." The profile went on to explain that "there are few, if any, major oil controversies in which Mr. Levy has not acted as a consultant," and that he "has been an advisor to most of the major oil companies, most of the important consuming countries and many of the large producing countries" ("As Oil Consultant, He's without Like or Equal," *New York Times*, July 27, 1969, sec. p. 3; Ed Shaffer, *The United States and the Control of World Oil* (New York: St. Martin's Press, 1983), 214–218).

76. Walter L. Buenger and Joseph A. Pratt, *But Also Good Business: Texas Commerce Banks and the Financing of Houston and Texas, 1886–1986* (College Station: Texas A&M University Press, 1986), 299.

77. Twentieth Century Fund Task Force on the International Oil Crisis, *Paying for Energy*, report (New York: McGraw-Hill, 1975), vii–viii; Twentieth Century Fund Task Force on United States Energy Policy, *Providing for Energy*, report (New York: McGraw-Hill, 1977), xi–xii; Robin W. Winks, *Laurence S. Rockefeller* (Washington DC: Island Press, 1997, 44 and 96). These task forces have the characteristics of economic elite-led policy discussion groups. These groups serve as means for economic elites and producer groups to come together with acceptable experts to form policy plans, and to develop a consensus on such plans and proposals (Barrow, *Critical Theories of the State*, chap. 1; Domhoff, *Who Rules America?*, chap. 4).

78. Twentieth Century Fund Task Force on the International Oil Crisis, *Paying for Energy*, 9.

79. Twentieth Century Fund Task Force on the International Oil Crisis, *Paying for Energy*, 9; emphasis in original.

80. Twentieth Century Fund Task Force on United States Energy Policy, *Providing for Energy*, 5; emphasis in original.

81. Twentieth Century Fund Task Force on the International Oil Crisis, *Paying for Energy*, 23; emphasis in original.

82. Twentieth Century Fund Task Force on United States Energy Policy, *Providing for Energy*, 23–24.

83. Twentieth Century Fund Task Force on United States Energy Policy, *Providing for Energy*, 24.

84. Twentieth Century Fund Task Force on the International Oil Crisis, *Paying for Energy*, 15.

85. Energy Information Administration (EIA), *Annual Energy Review 2003* (Washington, DC: U.S. Department of Energy, 2004), 57.

86. Rutledge, *Addicted to Oil*, 10.

87. According to the U.S. Energy Information Administration, global petroleum production in 2005 was 83.7 barrels per day. According to the United States Department of Energy, in 2005 the United States consumed about 9 million barrels per day of petroleum on powering automobiles (including light trucks and motorcycles). (United States Department of Energy, *Transportation Energy Data Book* (Washington, D.C.: United States Department of Energy, 2007), Table 2.6; Energy Information Administration, *Short-Term Energy Outlook* (Washington, D.C.: U.S. Department of Energy, 2008), Figure 5).

88. Duffield, 2008, chap. 2.

89. In 2005 U.S. petroleum consumption was 20.8 million barrels per day, whereas global petroleum production in 2005 was 83.7 million barrels per day. Sixty-seven percent of U.S. oil consumption in 2005 (13.9 million barrels per day) went toward transportation: automobiles, buses, aircrafts, shipping, and railways. Forty-three percent (8.9 million barrels per day) went toward producing "motor gasoline." Another 2.9 million barrels per day went toward producing distillate fuel oils, which includes diesel fuel. Diesel fuel is used to power private automobiles and buses, as well as heavy and medium trucks (Energy Information Administration, *Annual Energy Review 2006* (Washington, D.C.: U.S. Department of Energy, 2007), Table 5.1; United States Department of Energy, *Transportation Energy Data Book* (Washington, D.C.: United States Department of Energy, 2007), Table 2.6; Energy Information Administration, *Short-Term Energy Outlook* (Washington, D.C.: U.S. Department of Energy, 2008), Figure 5; Duffield, *Over of Barrel*, 19.

90. Philip, *The Political Economy of International Oil*, 195; Rutledge, *Addicted to Oil*, chap. 1; Robert Vandenbosch and Susanne E. Vandenbosch, *Nuclear Waste Stalemate: Political and Scientific Controversies* (Salt Lake City: University of Utah Press, 2007); Duffield, *Over a Barrel*, chap. 2.

91. Kenworthy and Laube et al., *An International Sourcebook of Automobile Dependence in Cities 1960–1990*.

92. Markku Lanne and Matti Liski, "Trends and Breaks in Per-Capita Carbon Dioxide Emissions, 1870–2028," *Energy Journal* 25, no. 4 (2004): 41–65; Kevin A. Baumert, Timothy Herzog, and Jonathan Pershing, *Navigating the Numbers: Greenhouse Gases and International Climate Change Agreements* (Washington, DC: World Resources Institute, 2005), 20; Germanwatch, The Climate Change Performance Index (Berlin: Germanwatch, 2007); International Energy Agency, CO_2 *Emissions From Fuel Combustion: 1971–2005* (Paris: International Energy Agency, 2007). In 2005 the United States, emitted 19.6 tons of carbon dioxide on a per capita, or per person, basis. This is in comparison to 10.8 tons emitted per person in Russia; 9.9 in Germany; 9.5 Japan; 9.3 South Korea; 8.8 United Kingdom; 6.2 France; 3.9 China; and 1.05 India. Six small countries in 2005 emitted more carbon dioxide than the United States on a per capita basis: Qatar,

Luxembourg, Netherlands Antilles, United Arab Emirates, Kuwait, and Bahrain (International Energy Agency, CO_2 *Emissions From Fuel Combustion*, Part Two, 49–51).

One body, the Netherlands Environmental Assessment Agency, holds that China in 2007 surpassed the United States as the world's largest emitter of carbon dioxide (Audra Ang, "China Overtakes U.S. As Top CO2 Emitter," Associated Press, June 21, 2007).

93. Olney, *Buy Now, Pay Later.*

94. Wolff, *Understanding Marx*; Michael Perelman, *The Perverse Economy* (New York: Palgrave Macmillan, 2003).

8

In the Wake of Katrina: Climate Change and the Coming Crisis of Displacement

Peter F. Cannavò

Some years ago, in an article on global warming, British economist Wilfred Beckerman made a rather astonishing argument. Contemplating the impacts and costs of sea-level rise, particularly on low-lying developing nations like Bangladesh, Beckerman argued that it would be pointless and expensive to try to prevent climate change and its effects. Beckerman maintained that "there would be very much cheaper ways of sparing them [i.e., Bangladeshis and other vulnerable populations] from these impacts." He suggested "helping them move away from the threatened coastal areas, building dikes . . . , improving flood control, and, perhaps allowing more of them to emigrate!"[1]

With some measure of climate change now inevitable, calls for some degree of adaptation certainly make sense. However, reliance on adaptation as *the* solution to global warming ignores the potential for unwelcome climatic surprises and truly catastrophic impacts. It also ignores the fact that, if left unchecked, climate change would be an open-ended process that could radically transform the biosphere and leave us with a very different planet to adapt to. Furthermore, appeals to adaptation and/or to future increased wealth and technological advances that will supposedly enable us to ride out global warming[2] ignore the often sheer ineptitude of human systems, and the all-too-frequent malfeasance of public officials, in dealing with catastrophe. The landfall of Hurricane Katrina on the Gulf Coast on August 29, 2005, and the subsequent devastation showed not only the potential for climatic disaster, but also the inability of government agencies, particularly the George W. Bush administration and the much-ridiculed Federal Emergency Management Agency, to react effectively and, truth be told, demonstrated the astounding capacity of public officials for callous, even malicious, negligence. While New Orleans drowned, President George W. Bush strummed a guitar at a naval base, posed for a photo op with Arizona Senator John

McCain and McCain's birthday cake, and spoke at various venues on the war in Iraq and the new Medicare prescription drug plan. Subsequent investigations, as well as the bitter recriminations leveled at the Bush administration and Department of Homeland Security Secretary Michael Chertoff by disgraced former FEMA head Michael Brown, only added to the picture of federal mismanagement. Meanwhile, the administrations of Louisiana Governor Kathleen Blanco and New Orleans Mayor Ray Nagin themselves earned few plaudits for competence.

However, what strikes me about Beckerman's remark is not so much the familiar call to adaptation, but the misguided assumption that whole populations can simply be moved out of danger. He views climate change as merely—I use that word advisedly—a problem of resources in which adaptation can be addressed with improved efficiency and distribution. In short, provide the means for populations to move elsewhere and they can escape environmental disaster relatively unscathed. Relatedly, Beckerman basically sees people as placeless—if it costs less to move than protect your home and your homeland, then start packing. Under this view, places and homes are little more than commodities—you can simply trade one for another if the price is right.

The Katrina disaster reveals the near absurdity of Beckerman's perspective. Total fatalities from Katrina are said to be at least 1,836; the damage has totaled more than $100 billion.[3] In New Orleans, the storm breached levees and flooded 80 percent of the city. Beyond the deaths and the sheer physical destruction was the loss of home and of place. According to one report, an estimated "700,000 or more people may have been acutely impacted by Hurricane Katrina, as a result of residing in areas that flooded or sustained significant structural damage."[4] The hurricane displaced, whether temporarily or permanently, about 1.5 million people.[5] Not only were homes literally destroyed, but places of deep personal and cultural attachment were ruined, perhaps irretrievably, and Gulf Coast residents were exiled to unfamiliar communities. Regarding Katrina and Hurricane Rita, which battered New Orleans and the Gulf Coast later in the season, the Brookings Institution reports that "the population dispersal they induced was the largest the United States has experienced during such a brief moment in time."[6]

The storm was itself part of a record-breaking Atlantic hurricane season. The 2005 season set records with twenty-seven named storms, fifteen hurricanes, four major hurricanes hitting the United States, and three Category 5 hurricanes.[7] Katrina may also be a harbinger of a global crisis of displacement and homelessness brought on by climate change.

The United Nations Institute for Environment and Human Security warns that by 2010 there may be over fifty million environmental refugees worldwide, with the number eventually growing into the hundreds of millions.[8] Even if the victims of intense hurricanes, rising sea levels, and other manifestations of global warming find new houses, neighbors, and jobs—an outcome that may be wildly optimistic—the loss to personal and collective identities and histories may be enormous. Offering Bangladeshis and others money and opportunities to move might be an unfortunate necessity in some circumstances, but it does not begin to address the enormity of the crisis.

The aim of this chapter is twofold. First, I wish to elaborate on the crisis of displacement that has followed in the wake of Hurricane Katrina. At issue is a loss of the fundamental values of "home" and "place." Second, I wish to elaborate on how climate change may force on us a tragic conflict between the value of place and another value with which it is frequently associated, ecological sustainability.

The Katrina Disaster

Katrina has been called the worst natural disaster in U.S. history. It devastated not just New Orleans, but also a wide swath of the Gulf Coast, including communities like Arabi, Bay St. Louis, Biloxi, Gulfport, Pass Christian, Port Sulphur, and Venice. Given the attention lavished on New Orleans, many of these communities have been relatively ignored by the national media. Here, I will admittedly continue that slight and focus on New Orleans. New Orleans' status as a major U.S. city with a distinctive and influential culture and identity and its location in precarious environmental conditions highlight in especially stark terms the sorts of dilemmas that may be posed by climate change.

In New Orleans, Katrina and its immediate aftermath provided graphic testimony to a supposedly advanced civilization's vulnerability to natural forces, as well as to America's deep racial and economic inequalities and political ineptitude. Thousands of mainly poor, minority, and older residents were trapped in New Orleans as the city was inundated. Looters roamed the streets, people were stranded on rooftops and highways, and evacuees faced appallingly inhumane conditions in the city's Superdome and Convention Center.

The victims of the storm tended to be among the most vulnerable or marginalized: "The 700,000 people acutely affected by Katrina were more likely than Americans overall to be poor; minority (most often

African-American); less likely to be connected to the workforce; and more likely to be educationally disadvantaged (i.e., not having completed a high school education)."[9] The areas in New Orleans hit hardest by the flooding were low-lying neighborhoods that tended to be African-American.[10] Moreover, according to the *New Orleans Times-Picayune*, data compiled by the State of Louisiana indicates that "more than 60 percent of Louisianans who died during Hurricane Katrina or immediately after were 61 or older. More than 37 percent were 76 or older."[11]

The city's recovery will be slow. Before Katrina, New Orleans had a population of 485,000. According to a March 2006 report by the RAND Corporation, the city's population was to have been at only 272,000, or about 56 percent, three years after Katrina.[12] In some ways, the recovery has been more rapid than RAND anticipated—two years after the storm, another Brookings report found New Orleans' population at 66 percent of its former level.[13] But this report also suggested cause for pessimism. According to the authors, New Orleans "continued to lose employers" during the year after Katrina, and also faced a rising unemployment rate. The authors found that "repairs to essential infrastructure [were] largely stalled and public services [were] still limited." There was continued housing repair and construction, but such rebuilding had slowed due to insufficient public funding. Meanwhile, "basic services—including schools, libraries, public transportation, and childcare—remain[ed] at less than half of the original capacity in New Orleans, and only two-thirds of all licensed hospitals [were] open in the region. Further, lack of repairs to public facilities [was] undermining police effectiveness."[14]

A major part of the recovery has entailed reversing a diaspora. The storm scattered New Orleanians across the nation, many as far as Maine or Washington State or even Alaska. Some New Orleanians, citing concern over vulnerability to hurricanes and the federal government's refusal to build levees strong enough to resist a Category 5 hurricane, plan on never moving back.[15] Many displaced residents do wish to come back, but the city they return to will be much changed. As people have moved back, the city's demographics have shifted significantly, indicating that serious, long-term dislocation has disproportionately burdened minority—especially African-American—and lower-income residents. The wealthy and white, many of whose higher-elevation neighborhoods were spared the flooding, have been more likely to return and rebuild; they were also better able to find temporary lodging closer to the city. During the year after Katrina, New Orleans went from a 67 percent black majority to one of 58 percent.[16] As for lower-income New Orleanians, they "migrated farther afield to places like Houston and Atlanta, and their status as renters, the greater devastation to their homes, or their precarious financial or labor force status may present obstacles to their short-term return to the region."[17]

Those scattered by Katrina often ended up in crowded shelters—an experience that for many began with a horrific stay in the New Orleans Superdome—or in hotels or trailers or other rental units and without jobs or income. Often, families were broken up, causing additional psychological stress, especially on children.[18] About 190,000 elementary and high school students were displaced by Katrina.[19] Children, in some cases separated from parents who managed to find work in New Orleans or other communities where schools were not available,[20] faced the combined challenges of living in new places under makeshift conditions and having to attend new schools. The National Fair Housing Alliance reported discrimination against black evacuees, in the form of higher rental rates than for whites or outright refusal to rent.[21] Though a number of communities and churches generously took in or arranged housing for Katrina evacuees,[22] in many cases the welcome was often uneasily accompanied by the hosts' fear of outsiders, dependency, and crime, and, perhaps, by racism.[23]

Dislocation

Dislocation has meant exactly that—a loss of location, of home, and of extended family, community, and neighborhood. This may be an especially hard predicament for New Orleanians: "One characteristic of New Orleans' population . . . is its strong 'rootedness.' That is, a much higher share of its residents were born in the same city and state than in other places, and therefore exhibited a strong attachment to the area."[24]

Seventy-one-year-old Gloria Jordan, having lost her house of forty-nine years and looking at the ruins of her porch, reminisced to an Associated Press reporter how "each morning she used to sip her coffee in a rocking chair and look out across her small garden of flowers." She told her, "Baby, I had a beautiful home. It's hard when you lived on your own for so many years and just like you pop your finger, or in the twinkling of an eye, you're homeless."[25]

Many New Orleanians are facing not just the loss of houses, but also the loss of place-based communities and social networks. Anna Mulrine writes that when residents do not return and rebuild, "the consequences are greater than that of a neighborhood's physical integrity." She quotes Louisiana State University sociologist Jeanne Hurlbert: "What makes this a catastrophe isn't just the loss of physical structures. It's the phenomenal destruction of networks, the enormous loss of emotional and social support." Among the casualties are family ties, which, says Hurlbert, are "what helped people hold down jobs and keep their kids safe."[26] Writing a few months after the storm, Mulrine described the plight of Peter and Sarah Parker:

For Peter Parker, the first step in the unraveling of his extended family's life happened in the blink of an eye, as he watched the floodwater "start to walk up the sewers" in front of his shotgun at Second Street and Freret. . . .

Eventually, the family was evacuated to San Antonio, but Sarah Parker was anxious to get home. Sitting on her front stoop surveying the empty shotgun houses that line her block, some days she wonders why. "People are gone," she says. "And most of them aren't coming back." Parker's sister used to live next door, but she decided to stay in San Antonio. Her house on Freret is being gutted, and the landlord is raising the rent once it's fixed up—too high to allow her to return. Parker's niece, who lived in the house next to that, is in Alabama "with her husband's people." No word on whether they're planning to return. Parker's husband, Peter, works about an hour and a half away, bunking on a couch with her relatives, returning to Freret Street every other weekend.

Lonely and depressed, Parker got counseling while she was in San Antonio, but she hasn't seen anyone since she came home. She sleeps most of the day, all four gas burners running on the stove to help keep the house warm. She used to work at Babykeepers Day Care Center, but the center is closed, and that job is gone. Plus, she says, now she has no one to look after her kids once they get home from school. . . . With the price for nearly everything skyrocketing and few signs that things are apt to get better on her block of Freret anytime soon, Parker understands the decisions of her neighbors. "If no one wants to come back," she says, "I'm not mad at em."[27]

The psychological impacts on those who lost homes, places, and communities have been enormous. In the June 21, 2006 *New York Times*, Susan Saulny reported, "New Orleans is experiencing what appears to be a near epidemic of depression and post-traumatic stress disorders, one that mental health experts say is of an intensity rarely seen in this country. It is contributing to a suicide rate that state and local officials describe as close to triple what it was before Hurricane Katrina struck and the levees broke 10 months ago." Saulny added, "Compounding the challenge, the local mental health system has suffered a near total collapse." She cited the sheer physical destruction of a home-place, along with the associated, ongoing social chaos and loss of livelihood, as key factors behind the spike in mental illness: "This is a city where thousands of people are living amid ruins that stretch for miles on end, where the vibrancy of life can be found only along the slivers of land next to the Mississippi. Garbage is piled up, the crime rate has soared, and as of Tuesday the National Guard and the state police were back in the city, patrolling streets that the Police Department has admitted it cannot handle on its own. The reminders of death are everywhere, and the emotional toll is now becoming clear." Gina Barbe, a resident who had worked in the tourism industry, remarked, "When I'm driving through the city, I have to pull to the side of the street and sob. I can't drive around this city without crying."[28]

A Unique City in Jeopardy

Ultimately, Katrina and its aftermath may fundamentally transform, and even diminish, New Orleans' character, particularly if many of the dislocated do not return. New Orleans pre-Katrina was certainly no utopia—writer Tom Piazza described it as "a city with enormous problems even on its best day."[29] The perhaps ironically named "Big Easy" has had a long history of racial inequality and polarization, huge disparities in wealth, and corrupt and inept government, characteristics that came out in force with Katrina.[30] Before Katrina, New Orleans had a murder rate ten times the U.S. average.[31] It also had unusually high rates of poverty. For example, according to the National Center for Children in Poverty at Columbia University, 38 percent of children in New Orleans were living at or below the poverty level, compared with 17 percent for the nation as a whole.[32]

Yet New Orleans, nicknamed the "Crescent City," has also been blessed with a rich, diverse, artistic, and often eccentric hybrid culture that is unique in the United States. The city has nurtured distinctive and enormously influential architectural, musical, culinary, literary, and festive traditions. Piazza says that "New Orleans inspires the kind of love that very few other cities do. Paris, maybe, Venice, maybe, San Francisco, New York . . . New Orleans has a mythology, a personality, a *soul*, that is large, and that has touched people around the world."[33]

The Crescent City's character grew out of its history of French and Spanish rule, its large African-American and Creole populations, and its waves of Caribbean and European immigration. When the city, founded in 1718, was acquired by the United States in 1803 as part of the

Louisiana Purchase, it was what Peirce Lewis describes as a "mature," foreign city.[34] The city's urbanity, architecture, Catholicism, decadence, and unusually complex race relations set it apart from the rest of the South.[35] Now New Orleans could lose much of its uniqueness.

A key issue is not only who has the wherewithal to return but also which neighborhoods actually get fully rebuilt. To help plan the city's reconstruction, Mayor C. Ray Nagin established the Bring New Orleans Back Commission (BNOBC). In a preliminary study commissioned by the BNOBC, the Urban Land Institute (ULI) raised questions about rebuilding neighborhoods that are especially flood-prone or have less chance of being significantly repopulated. The arguments were that flood-prone areas were unsafe until much more effective defenses against a Category 5 hurricane were built, and that it would make little economic sense to restore infrastructure and services to sparsely populated areas of the city.[36] Sparsely populated areas would also experience what is known as the "jack-o'-lantern syndrome," which Jed Horne describes as "the gap-toothed look of neighborhoods reviving unevenly," where there is "scattered rebuilding and widespread abandonment."[37] The ULI recommended that the city not rebuild infrastructure in such problematic neighborhoods and instead buy out the property owners. A January 2006 report by the BNOBC envisioned a more geographically compact city, with the historic districts preserved. In those areas "where the [flood] waters had been deepest, the blueprint called for reversion to green space, an archipelago of parks and retention ponds linked by pedestrian malls and bicycle paths." Areas that could not demonstrate an actual or anticipated 50 percent return rate for residents by mid-May would be bulldozed, and the residents bought out.[38] A conversion of floodplain neighborhoods to parkland had occurred in Grand Forks, North Dakota, after the city was flooded in 1997. However, the BNOBC's plan to contract the city's footprint was largely a nonstarter. Some New Orleanians had already begun to rebuild in the problematic areas. Moreover, the BNOBC's recommendations for not rebuilding disproportionately affected African-American neighborhoods, prompting charges of racism, suspicions of a land grab, and vows of resistance from property owners.[39] Journalist Bret Schulte reported that "the [May] deadline has sparked outrage and ignited a race against the clock in the black community to get residents to return. While networks of friends and families are reaching out across the country, grass-roots rebuilding efforts are springing up in black neighborhoods, encouraged by a majority-black City Council, which has repeatedly argued that the whole city should be redeveloped."[40]

Such concerns reflected black New Orleanians' fears about not only the loss of their homes and neighborhoods but also about the possible loss of their majority status in the city and the loss of New Orleans' distinctive culture, which is heavily indebted to the city's black population.[41] One African-American community activist, Lauren Anderson, said, "We have a culture here that's more intact than anything else in this country. As an African-American, I feel closer to my culture here than anywhere else in the country. And if we cannot reverse the diaspora, we'll lose traditions that are almost as old as this country itself."[42] New Orleans' African-American community has given the Crescent City its jazz heritage and many of its Mardi Gras traditions.[43] With a major demographic shift, the city could become a commodified ghost of its former self. In a widely quoted speech, Tulane University historian Lawrence N. Powell asked, "Will this quirky and endlessly fascinating place become an X-rated theme park, a Disneyland for adults?"[44] Stephen Bradberry, the head of ACORN in New Orleans, believes that "New Orleans without black people will be Disney World on the river."[45] Such concerns prompted not only the City Council, but also Nagin himself, to disavow many of the BNOBC recommendations. Nagin pledged not to cordon off any part of the city,[46] and he declared, "It is our intention to rebuild all of New Orleans."[47] He also made his now-infamous "chocolate city" remark: "It's time for us to rebuild a New Orleans, the one that should be a chocolate New Orleans."[48]

The Big Easy's Tenuous Existence

Horne remarks that Nagin's decision not to endorse a plan for shrinking the city's physical footprint "could be called a radical deference to democracy and the marketplace, or a massive default of leadership at a time of great trial."[49] As problematic as it was, the BNOBC report pointed to an undeniable truth, that New Orleans, or at least parts of it, are in an environmentally tenuous situation.

Katrina certainly was not New Orleans' first hurricane, and this was certainly not the first time the city had been flooded. Moreover, New Orleans' levee system had long been considered inadequate; indeed, as Horne documents, post-Katrina investigations discovered extremely shoddy construction and poor planning in the flood-protection system built by the Army Corps of Engineers.[50] Much of New Orleans is below sea level and southeastern Louisiana has been sinking, due to natural factors, drainage of lands for development, and petroleum pumping;

moreover, the Louisiana coastline has been losing wetlands and shrinking to the tune of twenty-five square miles per year as a result of development and canal construction:[51] "The canals cut into the Delta for navigation and to float oil-drilling platforms out to the Gulf disrupted the native vegetation by enabling salt or brackish water to penetrate deep into freshwater marshes."[52] This is especially problematic for New Orleans, as wetlands serve as an essential buffer against storm surges.

The Mississippi River Gulf Outlet is a premier example of how artificial factors have increased New Orleans' vulnerability to hurricanes and flooding. Starting in the 1920s, engineers worked to straighten the Mississippi River at the Delta to facilitate shipping, prevent New Orleans from being cut off from the Mississippi by natural changes in the river's course, and provide flood control. This effort eventually led to the creation of the Gulf Outlet, known locally as MR-GO and pronounced "Mister Go."

The geography of the Mississippi Delta is a dynamic interplay of land and water. The Mississippi periodically carves out new channels. The slow-moving waters of these channels deposit silt, which creates new land, or "lobes," and replenishes existing lobes. Eventually, the existing lobes recede, even as new lobes are created. Right now, the Mississippi is building up the Atchafalaya lobe.[53] However, MR-GO has retarded this buildup and depleted wetlands by greatly reducing the silt deposits that replenish coastal areas. The reason is that the artificial channel rapidly carries silt out to sea when it would otherwise be deposited by slower-moving natural waterways. Moreover, MR-GO has "acted like a funnel for storm surge," and carried floodwaters into the city.[54] According to one geologist, MR-GO has become "the icon of environmental evil in the region."[55]

New Orleans has always had an ambiguous relationship with the water that was the city's undoing. The Crescent City exists because of its economically and militarily strategic location at the mouth of the Mississippi. The city has flourished in large part because of shipping,[56] and the city's cuisine, typified by crawfish, shrimp, and seafood gumbo, reflects the Big Easy's debt to the water. Yet Lewis calls New Orleans "impossible but inevitable": the city's geographic location was quite favorable but the site—the swamp of the Mississippi Delta—was nearly inhospitable.[57] Consequently, the city had to be engineered into existence, through levees, drainage canals, and pumping stations that would keep water out of a constricted, bowl-shaped site that was largely below sea level and dangerously below the level of the Mississippi itself. Lewis notes that "if a city's situation is good enough, its site will be altered to make do."[58] The laid-back Big Easy is thus ironically an embodiment of humanity's attempt to conquer nature, an attempt that has always yielded partial success at best.[59] Location at the mouth of the Mississippi "guaranteed prosperity for New Orleans, but the site also guaranteed that the city would be plagued by incessant trouble: yellow fever, floods, and unbearable summer heat."[60]

Thus, despite water being the raison d'être for the existence of New Orleans, it has also been a source of fear. Rosemary James writes that "when New Orleanians are not in the midst of a disaster made by water, they generally prefer to forget that water and its dangers exist, turning their backs on some of the most gorgeous water views." She notes that "views of the Mississippi River from residences or restaurants are few and far between."[61] In many ways, the place that New Orleanians call home has rested on a tenuous containment of nature's destructive elements, a situation made even more insecure by human mismanagement and now, increasingly, by the problem of climate change.

The Katrina disaster may be a harbinger of future destructive events associated with climate change. Though we seem to be in a natural cycle of increased hurricane activity, there is also evidence that global warming may be boosting the intensity of hurricanes; such a trend, coupled with increased populations in coastal areas, would mean much more devastating hurricanes in this century.[62] The destructive impacts of climate change could be further magnified by rising sea levels. In fact, the Mississippi Delta is one of six deltas worldwide that researchers have identified as especially susceptible to sea-level rise over the next fifty years.[63]

The United Nations–sponsored Intergovernmental Panel on Climate Change (IPCC) reports that global average sea level rose about 0.17 meters during the twentieth century, and that sea-level rise accelerated—if we consider the period 1961 to 2003, the rate of sea-level rise was actually higher between 1993 and 2003 than over the entire period.[64] Based on a range of possible greenhouse gas emission scenarios, the IPCC estimates a 0.18 to 0.59 meter rise in global average sea level, as a result of thermal expansion of the oceans and melting of ice caps and glaciers, between 1980–1999 and 2090–2099.[65] This scenario will be worsened if ice caps and glaciers melt more rapidly than expected. Already, there is evidence of unprecedented and faster-than-expected thawing of ice sheets in Antarctica and Greenland, a trend highlighted by the spectacular collapse of Antarctica's Larsen-B ice shelf in 2002.[66]

The Significance of *Home*

The experience of Katrina shows that the impacts of climate change could include not only sheer physical destruction and death, but could also mean serious psychological scars on survivors. The loss endured by surviving New Orleanians was enormous. They lost their homes, both in terms of personal residence and in terms of neighborhood and city. For the many New Orleanians who are not returning, the loss of home in the Crescent City is permanent.

The concept of *home* signifies a place, or places, where one feels a special familiarity and security, a sense of reassurance and fit with one's identity, and even some measure of control over the environment or, at least, some degree of manageable predictability.[67] Home is, ideally, one's most familiar, legible, predictable, and safe place, a point of relative stability for an ever-changing person in a world always in flux. For Iris Marion Young, "home carries a core positive meaning as the material anchor for a sense of agency and a shifting and fluid identity."[68]

Home is often, though not invariably, one's own domicile. Home can also expand beyond one's domicile to embrace one's neighborhood, city, or region. We thus speak of one's homeland or, on a smaller geographical scale, home ground. One's homeland or home ground is marked by familiar locales and sustaining social networks and cultural norms and practices. Certainly, though, as "home" expands beyond one's own four walls, it takes in more difference, change, and complexity.[69] One of the distinctive marks of New Orleans has certainly been its rich, diverse street life, which is often evoked in post-Katrina memories:

I walked the Quarter with my friends, listening to music. It was cold and misty and the streets still glistened with diamondlike droplets from the earlier storm. One friend suggested beignets at Café du Monde to cap the evening. We cut down Pirate's Alley, which runs from Royal to Jackson Square, the heart of the old Creole town. . . .[70]

There is an extremely active street life and sidewalk life; people like to sit on their porches and talk, stop friends on street corners for a chat, unexpected parades. The streets themselves are small masterpieces of intrigue and grace, with names that invite you in—Story, Music, Desire, Harmony—and each has its own distinct personality.[71]

A homeland or home ground is an object of individual and collective identification and attachment. Journalist Rob Long remarks that "as anyone who's ever been to New Orleans will tell you, the people there were proud of declaring that they were their own special thing, their own special place. 'This isn't America,' someone told me when I was there two months ago, 'this is *New Orleans*.' He meant it, too."[72]

Geographer Yi-Fu Tuan asserts that "profound attachment to the homeland appears to be a worldwide phenomenon. It is not limited to any particular culture and economy. It is known to literate and non-literate peoples, hunter-gatherers, and sedentary farmers, as well as city dwellers." Such a place "nourishes."[73] Attachment to homeland or home ground may be "of a deep though subconscious sort [and] may come simply with familiarity and ease, with the assurance of nurture and security, with the memory of sounds and smells, of communal activities and homely pleasures accumulated over time."[74]

What helps knit one's personal domicile with its immediately sur-rounding home ground is what sociologists John Logan and Harvey Molotch call the "daily round." They write, "The place of residence is a focal point for the wider routine in which one's concrete daily needs are satisfied. . . . Defining a daily round is gradually accomplished as residents learn about needed facilities, their exact locations and offerings, and how taking advantage of one can be efficiently integrated into a routine that includes taking advantage of others." Dependence on the daily round and the locales in which it is enacted can also make one especially vulnerable to disruptions. In words that might apply to the experience of Hurricane Katrina, Logan and Molotch warn: "The devel-opment of an effective array of goods and services within reach of resi-dence is a fragile accomplishment; its disruption, either by the loss of one of the elements or by the loss of the residential starting place, can exact a severe penalty."[75] The loss experienced by New Orleanians involved the loss of entire neighborhoods, with their shops, workplaces, routines, and social networks. That New Orleanians have been relatively rooted as a population would make this loss even more jarring.

Home, Place, and Nature

In her discussion of homemaking, the practice of creating and maintain-ing a home, Young highlights the interactive relationship between found-ing, or creating, and preservation, or maintenance.[76] As I argue elsewhere,[77] *all* places—not just homes—are animated by an interaction, interdepen-dence, and tension between founding and preservation. We create places—whether through mapping or physical action—and then rely on their stability and familiarity. Stability and familiarity are never abso-lute—homes and other places evolve over time, often in response to our own changing needs and activities—though too much flux can render a place alien or inhospitable.

Place is often associated with ecological values. The argument is that attachment to place can lead to sustainable use or preservationist care for the land. Kirkpatrick Sale emphasizes "fully knowing the character of the natural world and being connected to it in a daily and physical way [that] provides [a] sense of oneness, of rootedness."[78] Val Plumwood speaks of "the deep and highly particularistic attachment to a place . . . based . . . on the formation of identity, social and personal, in relation to particular areas of land, yielding ties often as special and powerful as those to kin, and which are equally expressed in very specific and local responsibilities of care."[79] Barry Lopez describes "a kind of local expertise, an intimacy with place." Such a "specific geographical understanding . . . resides with men and women more or less sworn to a place, who abide there, who have a feel for the soil and the history."[80] In regard to environmental problems, Mark Sagoff says that "much of what we deplore about the human subversion of nature—and fear about the destruction of the environment—has to do with the loss of places we keep in shared memory and cherish with instinctive and collective loyalty."[81] Even green political theorists who have challenged the assumptions of the environmental movement have embraced connection to place as a basis for ecological responsibility.[82]

In truth, however, there is tension between the value of place and ecological values. Place, with its founding aspect, fundamentally involves a transformation of what is naturally given. When this transformation is a matter of merely mapping and naming the landscape, this is less of an issue. However, any act of physical cultivation or construction changes the natural landscape. The creation of useful places through physical transformation is necessary for human beings to inhabit the world, and the creation of such places, it must be acknowledged, represents a kind of victory over natural forces. Hannah Arendt famously articulates this view in *The Human Condition*.[83] Nature, she says, is characterized by relentless cycles of birth, growth, and decay. The natural world ultimately consumes all its creations: "Life is a process that everywhere uses up durability, wears it down, makes it disappear."[84] Human beings are in a "constant, unending fight against the processes of growth and decay."[85] Through the activity of work, human beings create enduring physical objects that can resist decay and other natural depredations and constitute a durable, built world, the "human artifice."[86]

The permanence of any cultivated field or built structure or settlement, or indeed any physical human artifact, is in defiance of natural wear and tear and decay, and the shelter and security associated with home and with human settlements are in part predicated on the ability of these artifacts to protect us against natural forces. Like other organisms, we need to modify our environment in order to survive in it, though with human beings, that modification is on an enormous scale. This is not at all to deny our interdependence with the rest of nature. In fact, the human artifice can best sustain itself through some degree of cooperation with natural forces rather than utter, unmitigated defiance. When a home or other place is not created and maintained through sustainable practices—when preservation is ignored—its very stability and security can prove false. A stable place thus represents a complex relationship with nature, a relationship combining elements of conquest and cooperation.

The Tragic Dilemma

Though Katrina has often been regarded as an "unnatural" disaster in that much of the blame lay with human incompetence, negligence, and malfeasance, the event has also revealed the degree to which place and home are fragile achievements in the face of natural forces. The flood-waters directly breached the security of cities, towns, and houses and even turned homes where people had sought refuge into death traps as the floodwaters rose. Moreover, after the floods receded, the waterlogged homes were claimed by natural forces of decay, as manifested by rampant mold infestations. The ravenous fungus had been kept at bay on the damp Gulf Coast by air conditioners and dehumidifiers. Horne quotes one New Orleans hotel owner: "In New Orleans, if you shut off power for thirty or forty days, it turns into a petri dish, no matter what."[87] In the wake of Katrina, mold overran abandoned homes and greeted return-ing residents with repulsive sights and a variety of health hazards.

At least until Katrina, the city of New Orleans had represented a kind of fragile victory over nature. But was this victory ever really sustainable? Did New Orleans represent a foolish attempt at pure conquest without sufficient cooperation with natural forces? This is too complex a question to be fully broached here. Yet, it is quite apparent that the creation and continued existence of New Orleans constituted an attempt to massively defy natural conditions, forces, and processes. It was hardly an exercise

in working with nature. That defiance made for very tenuous meteorological and hydrological conditions within which the city could survive.
That is why New Orleanians have been afraid of the water that has long
been "the source of both their charge and impending disaster."[88]

Aided by human failings, Katrina overwhelmed the tenuous conditions
on which the city relied. Climate change or not, it may be that New
Orleans was always doomed. Indeed, one might argue that New Orleans
should never have been built in the first place, that it is an environmental
fiasco that could never really provide a stable, secure, enduring home for
its residents. Certainly, however, the depletion of wetlands that naturally
protected the city against flooding and the construction of channels like
MR-GO vastly increased New Orleans' ecological untenability.

In the wake of Katrina, Klaus Jacob, a geophysicist and risk analyst,
expressed a decidedly unsentimental view of the Crescent City: "We have
always used New Orleans as the perfect example of the unsustainable
city. It is a hopeless case."[89] His recommendation for New Orleans was
dire: "It is time to face up to some geological realities and start a carefully planned deconstruction of New Orleans, assessing what can or
needs to be preserved, or vertically raised and, if affordable, by how
much. Some of New Orleans could be transformed into a 'floating city'
using platforms not unlike the oil platforms offshore, or, over the short
term, into a city of boathouses, to allow floods to fill in the 'bowl' with
fresh sediment."[90]

In a somewhat more optimistic vein, hydrologist Richard E. Sparks
called for rebuilding New Orleans in a way that would work with nature
rather than against it, what he called a "natural option."[91] A redesigned
New Orleans, he said, must take into account the potential for future
flooding as well as the predilection of the Mississippi to switch courses
within the Delta. Building better levees would not be enough. "New
Orleans will certainly be rebuilt," Sparks said,[92] but changes would have
to be made:

Looking at the recent flooding as a problem that can be fixed by simply strength-
ening levees will squander the enormous economic investment required and,
worse, put people back in harm's way. Rather, planners should look to science
to guide the rebuilding, and scientists now advise that the most sensible strategy
is to work with the forces of nature rather than trying to overpower them. This
approach will mean letting the Mississippi River shift most of its flow to a route
that the river really wants to take; protecting the highest parts of the city from
flooding and hurricane-generated storm surges while retreating from the lowest
parts; and building a new port city on higher ground that the Mississippi is
already forming through natural processes.[93]

Sparks envisioned the new city being constructed on the Atchafalaya lobe. He maintained that "'old' New Orleans would remain a national historic and cultural treasure, and continue to be a tourist destination and convention city. Its highest grounds would continue to be protected by a series of strengthened levees and other flood-control measures."[94] He added that "city planners and the government agencies (including FEMA) that provide funding for rebuilding must ensure that not all of the high ground is simply usurped for developments with the highest revenue return, such as convention centers, hotels, and casinos. The high ground also should include housing for the service workers and their families, so they are not consigned again to the lowest-lying, flood-prone areas."[95] This would mean safer ground for residents of lower-lying areas but it would also mean the razing of their existing neighborhoods, as happened in Grand Forks: "The flood-prone areas below sea level should be converted to parks and planted with flood-tolerant vegetation. If necessary, these areas would be allowed to flood temporarily during storms."[96]

Sparks's proposal, though perhaps ecologically sustainable and certainly less dire than Jacob's, would have still meant the end of New Orleans as we have known it. It would have erased many of the communities that generated the culture of "old" New Orleans. Even with the provision of affordable housing in a "new" New Orleans, old neighborhood ties and social networks would have been lost for good.

Such radical plans for contracting the city have not been adopted, and Nagin, as we saw, decided not to go along with the BNOBC recommendations. Nagin's decision may ultimately prove unwise—rebuilding the entire city may represent a self-defeating defiance of natural inevitability and a continuation of destructive attempts to reengineer nature. But it is not so easy to dismiss the stance of the mayor and other New Orleanians. New Orleanians, especially the city's black residents, are articulating attachments to the houses, neighborhoods, and city they have called home and to a distinctive culture in which they have participated. New Orleans resident James expresses this tenacity when she says of her city and its people, "Now we must find the inspiration for a new unfamiliar role: we must reinvent ourselves as heroes, capable of not only bringing a great city back, but of making it better for all New Orleanians; capable of facing down that old demon water and overcoming it with skill and ingenuity; capable of preserving the precious originality of New Orleans neighborhoods and their style, the music of New Orleans, its food, its lingo, its soul."[97] Given the ecological challenges, the security and stability of homes and home grounds in many New Orleans neighborhoods may have been illusory, and thus these places may have fallen short of the ideal of home. However, on the time scale of individual lives these places may have been stable and secure enough— even despite events like the Great Mississippi Flood of 1927 and Hurricane Betsy in 1965—to feel like home to the residents of a very rooted city.

And therein lies the tragic dilemma. Whether New Orleans ought to have been built in the first place or even that it will most probably experience serious flooding again does not automatically answer the question of whether it should continue to exist in its present form. Environmentally unsustainable or not, a truly stable and secure home or not, the city is a distinctive place to which its residents and much of the nation as a whole have become deeply attached. The value of place, so often allied with environmental values, here collides with them.

New Orleans may have long been a catastrophe in the making, and Katrina perhaps had little to do with climate change, but as the earth warms and the ecological impacts proliferate, a lot of other locales may find themselves facing a similar dilemma between the values of place and ecological sustainability. Global warming will overwhelm the complex relationships with nature underlying the existence of many more homes and homelands. I do not just mean coastal communities in the Mississippi Delta and other low-lying regions, but also communities where rainfall is scarce or excessive or where the temperature is often too hot or where potable water supplies depend on a winter snowpack or where subsistence hunters seek game on frozen seas or where sewer systems are vulnerable to intrusion from rising waters or where homes are at risk from forest fires or mudslides. . . . In other words, not just the more notoriously unsustainable places like New Orleans but countless places around the world, each with its collective life and social bonds, its history, its homes, its culture. Many of these places have, despite their ecological constraints, enjoyed relatively stable and hospitable natural conditions over the course of recent memory. Now, their continued existence as places will suddenly become ecologically problematic or even untenable.

Coping with climate change and not causing further damage by trying to defy ecological trends may require letting altered natural conditions take their course and erase homes, homelands, and cultural treasures. We cannot simply defy a changing climate and hang on to places that are increasingly threatened with destruction; we must adapt.

Some adaptation is inevitable, but reliance on adaptation as a solution to global warming would mean that many more places will be endangered. We are better off trying to mitigate climate change and minimize the tragic dilemmas by reducing fossil fuel consumption and deforestation.

If old New Orleans and other homes and homelands are lost, the crisis of displacement and its tragic dilemma will play itself out on many levels. In a March 1, 2005, segment on National Public Radio's *All Things Considered*, Torrie Lawson, a displaced New Orleans junior high school student, told interviewer Michele Norris, "If you had told me I would have lived in a trailer before Katrina, I would have took it as an insult. I wouldn't have believed you. But now, it's like—it's nothing like my home. I wish I could be home *so* much, in my house."

Notes

I would like to thank Bill Chaloupka and John Barry for their comments on this chapter. The chapter was originally presented as a paper at the Western Political Science Association Annual Meeting, on March 16, 2006, in Albuquerque, New Mexico, as part of the panel, "Political Theory in the Greenhouse I: Theorizing Responses to Disturbance, Displacement, and Disaster." Portions of this chapter also appear in my book, *The Working Landscape: Founding, Preservation, and the Politics of Place* (Cambridge, MA: MIT Press, 2007), and are reprinted by permission of the MIT Press.

1. Wilfred Beckerman, "Global Warming and International Action: An Economic Perspective," in Andrew Hurrell and Benedict Kingsbury, eds., *The International Politics of the Environment* (Oxford: Oxford University Press, 1992), 253–289 (267). The exclamation point is in the original text.

2. For this sort of argument, see also Bjørn Lomborg, *The Skeptical Environmentalist: Measuring the Real State of the World*, rpt. ed. (Cambridge: Cambridge University Press, 2001).

3. Emily Harrison, "Suffering a Slow Recovery," *Scientific American*, 297, no. 3 (September 2007), 22–25; "Politics May Doom New Orleans" (editorial), *San Francisco Chronicle*, August 25, 2006, B10; "The Long Slog After Katrina" (editorial), *Boston Globe*, August 29, 2007, A16.

4. Thomas Gabe, Gene Falk, and Maggie McCarty, *Hurricane Katrina: Social-Demographic Characteristics of Impacted Areas* (Washington, DC: Congressional Research Service, 2005), ii.

5. Harrison, "Suffering a Slow Recovery," 14.

6. William H. Frey and Audrey Singer, *Katrina and Rita Impacts on Gulf Coast Populations: First Census Findings* (Washington, DC: Brookings Institution, 2006), 15.

7. U.S. Department of Commerce / National Oceanic and Atmospheric Administration, "NOAA Reviews Record-Setting 2005 Atlantic Hurricane Season," www.noaanews.noaa.gov/stories2005/s2540.htm.

8. United Nations University, Institute for Environment and Human Security, "As Ranks of 'Environmental Refugees' Swell Worldwide, Calls Grow for Better Definition, Recognition, Support" (press release), October 12, 2005, http://www.ehs.unu.edu.

9. Gabe, Falk, and McCarty, *Hurricane Katrina*, 13.

10. On the reasons for the relationship between the distribution of the city's black population and low elevation and the connections with racial discrimination, see Peirce F. Lewis, *New Orleans: The Making of an Urban Landscape* (Santa Fe, NM: Center for American Places, 2003), 50–52, and Craig E. Colten, *An Unnatural Metropolis: Wresting New Orleans From Nature* (Baton Rouge: Louisiana State University Press, 2005), 77–107.

11. Jarvis DeBerry, "When Life's Haven Turns Deathtrap," *New Orleans Times-Picayune*, November 6, 2005, 7 (Metro Section).

12. Kevin McCarthy, D. J. Peterson, Narayan Sastry, and Michael Pollard, *The Repopulation of New Orleans After Hurricane Katrina* (Santa Monica: RAND Corporation, 2006).

13. Amy Liu and Allison Plyer, *New Orleans Index, Second Anniversary Edition: A Review of Key Indicators of Recovery Two Years After Katrina* (Washington, DC: Brookings Institution and Greater New Orleans Community Data Center, 2007).

14. Liu and Plyer, *New Orleans Index*, 1.

15. Anna Mulrine, "The Long Road Back," *U.S. News & World Report*, February 27, 2006, 44–50, 52–58.

16. William H. Frey, Audrey Singer, and David Park, *Resettling New Orleans: The First Full Picture from the Census* (Washington, DC: Brookings Institution, 2007).

17. Frey and Singer, *Katrina and Rita Impacts on Gulf Coast Populations*, 11.

18. For the story of one high school student and his family, see Jere Longman, "From the Ruins of Hurricane Katrina Come a Championship and Concerns," *New York Times*, January 22, 2006, 1 (sec. 8).

19. Linda Jacobson, "Hurricanes' Aftermath Is Ongoing," *Education Week* 25, no.21 (February 1, 2006): 1–18.

20. See Jacobson, "Hurricanes' Aftermath Is Ongoing"; Merri Rosenberg, "Displaced by Katrina, Coping in a New School," *New York Times*, December 18, 2005, 2 (Westchester Weekly Section); Emma Daly, "Helping Students Cope with a Katrina-Tossed World," *New York Times*, November 16, 2005, B9.

21. Sarah Childress, "Welcome: 'White Couple,'" *Newsweek*, January 9, 2006, 8.

22. See, for example, Shirley Henderson, "After the Hurricanes: Opening Hearts & Homes to the Evacuees, *Ebony*, December 2005, 162.

23. See Katherine Boo, "Shelter and the Storm," *New Yorker* 81, no. 38 (November 28, 2005): 82–97.

24. Frey and Singer, *Katrina and Rita Impacts on Gulf Coast Populations*, 11.

25. Rukmini Callimachi, "Half a Year Later, New Orleans Is Far from Whole," Associated Press State & Local Wire, February 22, 2006. See also Willmarine Hurst, "Longing for Home and How It Used to Be," *New Orleans Times-Picayune*, November 14, 2005, 5 (Metro Section).

26. Mulrine, "The Long Road Back."

27. Mulrine, "The Long Road Back."

28. Susan Saulny, "A Legacy of the Storm: Depression and Suicide," *New York Times*, June 21, 2006, A1.

29. Tom Piazza, *Why New Orleans Matters* (New York: HarperCollins, 2005), xix–xx.

30. For an account of the performance of the city's Police Department, see Dan Baum, "Deluged," *New Yorker* 81, no.43 (January 9, 2006): 50–63. "As an institution," Baum says, "the New Orleans Police Department disintegrated with the first drop of floodwater."

31. Baum, "Deluged."

32. National Center for Children in Poverty, Columbia University, "Child Poverty in States Hit by Hurricane Katrina," Fact Sheet No.1, September 2005, www.nccp.org/pub_cpt05a.html.

33. Piazza, *Why New Orleans Matters*, xviii; emphasis in original.

34. Lewis, *New Orleans*, 5.

35. Lewis, *New Orleans*, 15–16.

36. Jed Horne, *Breach of Faith: Hurricane Katrina and the Near Death of a Great American City* (New York: Random House, 2006), 316.

37. Horne, *Breach of Faith*, 318.

38. Horne, *Breach of Faith*, 323–324.

39. Horne, *Breach of Faith*, 316, 321, 324. See also Gary Rivlin, "Wealthy Blacks Oppose Plans for Their Property," *New York Times*, December 10, 2005, A11.

40. Bret Schulte, "Turf Wars in the Delta," *U.S. News & World Report* 140, no. 7 (February 27, 2006): 66–71.

41. Manuel Roig-Franzia, "A City Fears for Its Soul: New Orleans Worries That Its Unique Culture May Be Lost," *Washington Post*, February 3, 2006, A1.

42. Mulrine, "The Long Road Back."

43. Roig-Franzia, "A City Fears for Its Soul."

44. Roig-Franzia, "A City Fears for Its Soul."

45. Schulte, "Turf Wars in the Delta."

46. Horne, *Breach of Faith*, 326.

47. Quoted in Horne, *Breach of Faith*, 320.

48. Quoted in Horne, *Breach of Faith*, 314.

49. Horne, *Breach of Faith*, 326.

50. Horne, *Breach of Faith*, 144–167, 327–340, 381–383.

51. Lewis, *New Orleans*, 22; Colten, An *Unnatural Metropolis*, 162–185; Jeff Hecht, "Six Deadly Deltas Eight Million People," *New Scientist*, February 18, 2006, 8; Juliet Eilperin, "Shrinking La. Coastline Contributes to Flooding," *Washington Post*, August 30, 2005, A7; Bob Marshall, "The River Wild: Rebuilding Our Coastal Wetlands the Old-Fashioned Way," *New Orleans Times-Picayune*, March 10, 2006, 7.

52. Richard E. Sparks, "Rethinking, Then Rebuilding New Orleans," *Issues in Science & Technology* 22, no. 2 (winter 2006): 33–39.

53. Sparks, "Rethinking, Then Rebuilding New Orleans."

54. Horne, *Breach of Faith*, 153.

55. Matthew Brown, "Reasons to Go," *New Orleans Times-Picayune*, January 8, 2006, 1.

56. Lewis, *New Orleans*, 8.

57. Lewis, *New Orleans*, 19–20.

58. Lewis, *New Orleans*, 20; see also Colten, *An Unnatural Metropolis*.

59. On the reengineering of the Mississippi Delta, see John McPhee, *The Control of Nature* (New York: Farrar, Straus and Giroux, 1990).

60. Lewis, *New Orleans*, 20.

61. Rosemary James, "New Orleans Is a Pousse-Café," in Rosemary James, ed., *My New Orleans: Ballads to the Big Easy by Her Sons, Daughters, and Lovers*, 1–20 (New York: Simon and Schuster, 2006) (quote on 3).

62. See Kerry Emanuel, "Increasing Destructiveness of Tropical Cyclones Over the Past 30 Years," *Nature* 436, no.7051 (August 4, 2005): 686–688.

63. Hecht, "Six Deadly Deltas, Eight Million People."

64. Intergovernmental Panel on Climate Change, *Climate Change 2007: The Physical Science Basis, Summary for Policymakers* (Geneva: World Meteorological Organization and the United Nations Environment Programme, February 2007), 5.

65. IPCC, *Climate Change 2007: Physical Science Basis*, 11.

66. Eugene Domack, Diana Duran, Amy Leventer, Scott Ishman, Sarah Doane, Scott McCallum, David Amblas, Jim Ring, Robert Gilbert, and Michael Prentice, "Stability of the Larsen B Ice Shelf on the Antarctic Peninsula during the Holocene Epoch," *Nature* 436, no. 7051 (August 4, 2005): 681–685; Eric Hand, "Antarctic Ice Loss Speeding Up," *NatureNews*, January 13, 2008 (www.nature .com/news/2008/080113/full/news.2008.438.html); Bob Holmes, "Melting Ice, Global Warning," *New Scientist*, October 2, 2004, 8; Fred Pearce, "The Flaw

in the Thaw," *New Scientist*, August 27, 2005, 26; Larry Rohter, "Antarctica, Warming, Looks Ever More Vulnerable," *New York Times*, January 25, 2005, F1.

67. See Peter F. Cannavò, *The Working Landscape: Founding, Preservation, and the Politics of Place* (Cambridge, MA: MIT Press, 2007), 25–30; Iris Marion Young, *Intersecting Voices: Dilemmas of Gender, Political Philosophy, and Policy* (Princeton, NJ: Princeton University Press, 1997), 134–164; Yi-Fu Tuan, *Topophilia: A Study of Environmental Perception, Attitudes, and Values* (New York: Columbia University Press, 1990), 99–100; Robert David Sack, *Homo Geographicus: A Framework for Action, Awareness, and Moral Concern* (Baltimore: Johns Hopkins University Press, 1997), 14–17.

68. Young, *Intersecting Voices*, 159.

69. See, for example, Iris Marion Young's discussion of urban life in *Justice and the Politics of Difference* (Princeton, NJ: Princeton University Press, 1990), 234–241.

70. James, "New Orleans Is a Pousse-Café," 11.

71. Piazza, *Why New Orleans Matters*, 15.

72. Rob Long, "Different Down There: In America, But Not Entirely," *National Review*, September 26, 2005, 32, 34 (quote on 32).

73. Yi-Fu Tuan, *Space and Place: The Perspective of Experience* (Minneapolis: University of Minnesota Press, 1977), 154.

74. Tuan, *Space and Place*, 159.

75. John Logan and Harvey Molotch, *Urban Fortunes: The Political Economy of Place* (Berkeley: University of California Press, 1988), 103–104.

76. Young, *Intersecting Voices*, 152.

77. Cannavò, *The Working Landscape*.

78. Kirkpatrick Sale, *Dwellers in the Land: The Bioregional Vision* (San Francisco: Sierra Club Books, 1985), 47.

79. Val Plumwood, "Nature, Self, and Gender: Feminism, Environmental Philosophy, and the Critique of Rationalism," in Michael Zimmerman, J. Baird Callicott, George Sessions, Karen J. Warren, and John Clark, eds., *Environmental Philosophy: From Animal Rights to Radical Ecology*, 284–309 (Englewood Cliffs: Prentice Hall, 1993) (quote on 297).

80. Barry Lopez, *About This Life: Journeys on the Threshold of Memory* (New York: Knopf, 1998), 132.

81. Mark Sagoff, "Settling America: The Concept of Place in Environmental Politics," in Philip D. Brick and R. McGreggor Cawley, eds., *A Wolf in the Garden: The Land Rights Movement and the New Environmental Debate*, 249–260 (Lanham, MD: 1996) (quote on 249).

82. See, for example, John M. Meyer, *Political Nature: Environmentalism and the Interpretation of Western Thought* (Cambridge, MA: MIT Press, 2001),

136–141; Andrew Biro, *Denaturalizing Ecological Politics: Alienation from Nature from Rousseau to the Frankfurt School and Beyond* (Toronto: University of Toronto Press, 2005), 210–211.

83. Hannah Arendt, *The Human Condition* (Chicago: University of Chicago Press, 1958).

84. Arendt, *The Human Condition*, 96.

85. Arendt, *The Human Condition*, 100.

86. Arendt, *The Human Condition*, 136.

87. Horne, *Breach of Faith*, 298.

88. James, "New Orleans Is a Pousse-Café," 4.

89. Quoted in Richard Ingham, "New Orleans Disaster Serves Up a Tough Lesson on Environment," Agence France-Presse, September 1, 2005, http://www.afp.com/english/globe/?pid=archives.

90. Klaus Jacob, "Time for a Tough Question: Why Rebuild?", *Washington Post*, September 6, 2005, A25.

91. Sparks, "Rethinking, Then Rebuilding New Orleans," 38.

92. Sparks, "Rethinking, Then Rebuilding New Orleans," 33.

93. Sparks, "Rethinking, Then Rebuilding New Orleans," 33.

94. Sparks, "Rethinking, Then Rebuilding New Orleans," 38.

95. Sparks, "Rethinking, Then Rebuilding New Orleans," 38.

96. Sparks, "Rethinking, Then Rebuilding New Orleans," 38.

97. James, "New Orleans Is a Pousse-Café," 19–20.

References

Abel, Tom. "Complex Adaptive Systems, Evolutionism, and Ecology within Anthropology: Interdisciplinary Research for Understanding Cultural and Ecological Dynamics." *Georgia Journal of Ecological Anthropology* 2 (1998): 6–29.

Abramson, Rudy. *Spanning the Century: The Life of W. Averell Harriman, 1891–1986.* New York: Morrow, 1992.

Aceves, William J. "Critical Jurisprudence and International Legal Scholarship: A Study of Equitable Distribution." *Columbia Journal of Transnational Law* 39 (2001): 299–393.

ACIA. *Impacts of a Warming Arctic: Arctic Climate Impact Assessment.* Cambridge: Cambridge University Press, 2004.

Adam, David. "Goodbye Sunshine." *The Guardian*, December 18, 2003.

Adger, Neil. "Social and Ecological Resilience: Are They Related?" *Progress in Human Geography* 24, no. 3 (2000): 347–364. Agarwal, Anil, and Sunita Narain. *Global Warming in an Unequal World: A Case of Environmental Colonialism.* New Delhi: Center for Science and Environment, 1991.

Agrawal, Arun. "Environmentality: Community, Intimate Government, and the Making of Environmental Subjects in Kumaon, India." *Current Anthropology* 46, no. 2 (2005): 161–190.

Agrawal, Arun. *Environmentality: Technologies of Government and the Making of Subjects.* Durham, NC: Duke University Press, 2005.

American Petroleum Institute. *Petroleum Facts and Figures.* New York: American Petroleum Institute, 1959.

Anderies, J. M., M. A. Janssen, and E. Ostrom. "A Framework to Analyze the Robustness of Social-Ecological Systems from an Institutional Perspective." *Ecology and Society* 9, no. 1 (2004): 18.

Andersen, A. N., G. D. Cook, and R. J. Williams, eds. *Fire in Tropical Savannas: the Kapalga Experiment.* New York: Springer, 2003.

Anderson, Soren, and Richard Newell. "Prospects for Carbon Capture and Storage Technologies." *Annual Review of Environment and Resources* 29 (2004): 109–142.

Anderson, Terry L., and Donald R. Leal. *Free Market Environmentalism*. New York: Palgrave, 2001.

Archer, David. "Fate of Fossil Fuel CO_2 in Geologic Time." *Journal of Geophysical Research* 110 (2005): C09S05.

Archer, David. "How Long Will Global Warming Last?" http://www.realclimate.org/index.php/archives/2005/03/how-long-will-global-warming-last/#more-134.

Arctic Council. *Arctic Marine Strategic Plan*. 2004. Available at www.pame.is.

Arendt, Hannah. *The Human Condition*. Chicago: University of Chicago Press, 1958.

Arno, Stephen F., and Steven Allison-Bunnell. *Flames in Our Forest: Disaster or Renewal*. Washington, DC: Island Press, 2002.

Arrhenius, Svante. "On the Influence of Carbonic Acid in the Air upon the Temperature of the Ground." *Philosophical Magazine* 41 (1896): 237–276.

Arrhenius, Svante. *Worlds in the Making: The Evolution of the Universe*. Trans. H. Borns. New York: Harper & Brothers, 1908.

"As Oil Consultant, He's without Like or Equal." *New York Times*, July 27, 1969.

Athanasiou, Thomas, and Paul Baer. *Dead Heat: Global Justice and Global Warming*. New York: Seven Stories Press, 2002.

Atkinson, Robert D. *The Past and Future of America's Economy*. Northampton, MA: Edward Elgar, 2004.

Attfield, Robin. *The Ethics of the Global Environment*. West Lafayette, IN: Purdue University Press, 1999.

Auburn, F. M. *Antarctic Law and Politics*. Bloomington: Indiana University Press, 1982.

Axelrod, Regina S., David Leonard Downie, and Norman J. Vig, eds. *The Global Environment: Institutions, Law, and Policy*. Washington, DC: CQ Press, 2005.

Bachram, Heidi. "Climate Fraud and Carbon Colonialism: The New Trade in Greenhouse Gases." *Capitalism, Nature, Socialism* 15 (2004): 5–20.

Baer, Paul. "Equity, Greenhouse Gas Emissions, and Global Common Resources." In Stephen H. Schneider, Armin Rosencranz, and John O. Niles, eds., *Climate Change Policy: A Survey*. Washington, DC: Island Press, 2002.

Barach, Arnold B., and the Twentieth Century Fund. *USA and Its Economic Future*. New York: Macmillan, 1964.

Bardou, Jean-Pierre, Jean-Jacques Chanaron, Patrick Fridenson, and James M. Laux. *The Automobile Revolution*. Chapel Hill: University of North Carolina Press, 1982.

Barnes, Peter. *Who Owns the Sky?* Washington, DC: Island Press, 2001.

Barrow, Clyde W. *Critical Theories of the State*. Madison: University of Wisconsin Press, 1993.

Barry, Brian. *Justice as Impartiality.* Oxford: Oxford University Press, 1995.

Barry, Brian. "Sustainability and Intergenerational Justice." In Andrew Dobson, ed., *Fairness and Futurity.* New York: Oxford University Press, 1999.

Baslar, Kernel. *The Concept of the Common Heritage of Mankind in International Law.* The Hague: Martinus Nijhoff, 1998.

Baudrillard, Jean. *For a Critique of the Political Economy of the Sign.* St. Louis, MO: Telos Press, 1981.

Baudrillard, Jean. *The Transparency of Evil: Essays on Extreme Phenomena.* London: Verso, 1993.

Baum, Dan. "Deluged." *New Yorker* 81, no. 43 (January 9, 2006): 50–63.

Baumert, Kevin A., Timothy Herzog, and Jonathan Pershing. *Navigating the Numbers: Greenhouse Gases and International Climate Change Agreements.* Washington, DC: World Resources Institute, 2005.

Beck, Peter J. *The International Politics of Antarctica.* London: Croom and Helm, 1986.

Beck, Ulrich. *What Is Globalization?* Oxford: Blackwell, 2000.

Beckerman, Wilfred. "Global Warming and International Action: An Economic Perspective." In Andrew Hurrell and Benedict Kingsbury, eds., *The International Politics of the Environment.* Oxford: Oxford University Press, 1992.

Beitz, Charles R. *Political Theory and International Relations.* Princeton, NJ: Princeton University Press, 1979.

Berkes, Fikret, and Carl Folke. "Investing in Cultural Capital for Sustainable Use of Natural Capital." In A. M. Jansson, M. Hammer, C. Folke, and R. Constanza, *Investing in Natural Capital: The Ecological Economics Approach to Sustainability.* Washington, DC: Island Press, 1994.

Betsill, Michele M. "Global Climate Change Policy: Making Progress or Spinning Wheels?" In Regina S. Axelrod, David Leonard Downie, and Norman J. Vig, *The Global Environment: Institutions, Law, and Policy.* Washington, DC: CQ Press, 2005.

Betsill, Michele M., Kathryn Hochstetler, and Dimitris Stevis, eds. *Palgrave Advances in International Environmental Politics.* Basingstoke, UK: Palgrave Macmillan, 2006.

Bina, Cyrus. *The Economics of the Oil Crisis.* New York: St. Martin's Press, 1985.

Bina, Cyrus. "Some Controversies in the Development of Rent Theory: The Nature of Oil Rent." *Capital & Class* 39 (1989): 82–112.

Biro, Andrew. *Denaturalizing Ecological Politics: Alienation from Nature from Rousseau to the Frankfurt School and Beyond.* Toronto: University of Toronto Press, 2005.

Blair, John M. *The Control of Oil.* New York: Pantheon, 1976.

Blatt, Harvey. *America's Environmental Report Card.* Cambridge, MA: MIT Press, 2005.

Bodansky, Daniel. "The Legitimacy of International Governance: A Coming Challenge for International Environmental Law?" *American Journal of International Law* 93, no. 3 (1999): 596–624.

Boo, Katherine. "Shelter and the Storm." *New Yorker* 81, no. 38 (2005): 82–97.

Bradsher, Keith. *High and Mighty: SUVs*. New York: Public Affairs, 2002.

British Broadcasting Corporation. "Global Dimming: Horizon producer David Sington answers questions about global dimming." January 13, 2005. http://www.bbc.co.uk/sn/tvradio/programmes/horizon/dimming_qa.shtml.

British Broadcasting Corporation. "Global Dimming." Program transcript. January 13, 2005. http://www.bbc.co.uk/sn/tvradio/programmes/horizon/dimming_trans.shtml.

Bromley, Daniel W. "Comment: Testing for Common versus Private Property." *Journal of Environmental Economics and Management* 21 (1991): 92–96.

Bronson, Rachel. *Thicker Than Oil: America's Uneasy Partnership with Saudi Arabia*. New York: Oxford University Press, 2006.

Broome, John. *Counting the Cost of Global Warming*. Isle of Harris, UK: White Horse Press, 1992.

Brosius, J. Peter. "Green Dots, Pink Hearts: Displacing Politics from the Malaysian Rain Forest." *American Anthropologist* 101 (1999): 36–57.

Brown, Donald A. *American Heat: Ethical Problems with the United States' Response to Global Warming*. Lanham, MD: Rowman & Littlefield, 2002.

Brown, Matthew. "Reasons to Go." *New Orleans Times-Picayune*, January 8, 2006, 1.

Browne, Marjorie A. *The Law of the Sea Convention and U.S. Policy*. Washington, DC: Congressional Research Service, 2000.

Bryant, Bunyan, and Paul Mohai, eds. *Race and the Incidence of Environmental Hazards: A Time for Discourse*. Boulder, CO: Westview Press, 1992.

Buenger, Walter L., and Joseph A. Pratt. *But Also Good Business: Texas Commerce Banks and the Financing of Houston and Texas, 1886–1986*. College Station: Texas A&M University Press, 1986.

Bullard, Robert D. *Dumping in Dixie: Race, Class, and Environmental Quality*. Boulder, CO: Westview Press, 1990.

Bullard, Robert D. "Waste and Racism: A Stacked Deck?" *Forum for Applied Research and Public Policy* 8 (1993): 29–45.

Burkett, Paul. *Marx and Nature*. New York: St. Martin's, 1999.

Burkett, Paul. "Nature's 'Free Gifts' and the Ecological Significance of Value." *Capital & Class* 68 (1999): 89–110.

Burkett, Paul. *Marxism and Ecological Economics: Toward a Red and Green Political Economy*. Boston: Brill, 2006.

Burtraw, Dallas, and Karen Palmer. *The Paparazzi Take a Look at a Living Legend: The SO_2 Cap-and-Trade Program for Power Plants in the United States*. Washington, DC: Resources for the Future, 2003.

Busenberg, George J. "Adaptive Policy Design for the Management of Wildfire Hazards." *American Behavioral Scientist* 48, no. 3 (2004): 314–326.

"Bush Administration Proposal for Reducing Greenhouse Gases." *American Journal of International Law* 96, no. 2 (2002): 488.

Callimachi, Rukmini. "Half a Year Later, New Orleans Is Far from Whole." Associated Press State & Local Wire, February 22, 2006.

Cannavò, Peter F. *The Working Landscape: Founding, Preservation, and the Politics of Place.* Cambridge, MA: MIT Press, 2007.

Capek, Stella M. "The 'Environmental Justice' Frame: A Conceptual Discussion and an Application." *Social Problems* 40, no. 1 (1993): 5–24.

Carle, David. *Burning Questions: America's Fight with Nature's Fire.* Westport, CT: Praeger, 2002.

Carskadon, Thomas Reynolds, and George Henry Soule. *USA in New Dimensions: The Measure and Promise of America's Resources, A Twentieth Century Fund Survey.* New York: Macmillan, 1957.

Cash, D. W., and S. C. Moser. "Linking Local and Global Scales: Designing Dynamic Assessment and Management Processes." *Global Environmental Change* 10 (2000): 109–120.

Centre for Science and Environment (Delhi). "The Leader of the Most Polluting Country in the World Claims Global Warming Treaty Is 'Unfair' Because It Excludes India and China." March 16, 2001. http://www.cseindia.org/html/au/au4_20010317.htm.

Chapin, F. S. III, A. L. Lovecraft, E. S. Zavaleta, J. Nelson, M. D. Robards, G. P. Kofinas, S. F. Trainor, G. Peterson, H. P. Huntington, and R. L. Naylor. "Policy Strategies to Address Sustainability of Alaskan Boreal Forests in Response to a Directionally Changing Climate." *Proceedings of the National Academy of Sciences: Early Edition.* 2006. www.pnas.org/cgi/doi/10.1073/pnas.0606955103.

Chapin, S. C., S. T. Rupp, A. M. Starfield, L. DeWilde, E. Zavaleta, N. Fresco, J. Henkelman, and D. A. McGuire. "Planning for Resilience: Modeling Change in Human-Fire Interactions in the Alaskan Boreal Forest." *Frontiers in Ecology* 1, no. 5 (2003): 255–261.

Chapin, S. C., B. H. Walker, R. J. Hobbs, D. U. Hooper, J. H. Lawton, O. E. Sala, and D. H. Tilman. "Biotic Control over the Functioning of Ecosystems." *Science* 277 (1997): 500–504.

Charlson, Robert J., Francisco P. J. Valero, and John H. Seinfeld. "Atmospheric Science: In Search of Balance." *Science* 308, no. 5728 (May 6, 2005): 806–807.

Chen, Jiyang, and Atsuma Ohmura. "Estimation of Alpine Glacier Water Resources and Their Change Since 1870s." *Hydrology in Mountainous Regions 1, IAHS Publication* 193 (1990): 127–135.

Childress, Sarah. "Welcome: 'White Couple.'" *Newsweek*, January 9, 2006, 8.

Christhoff, Peter. "Weird Weather and Climate Culture Wars." *ARENA Journal* 23 (2005): 9–17.

Clark, William C., and Nancy Dickson. "Sustainability Science: The Emerging Research Program." *Proceedings of the National Academy of Sciences* 100, no. 14 (2003): 8059–8061.

Claussen, Eileen., and Lisa McNeilly. *The Complex Elements of Global Fairness.* Washington, DC: Pew Center on Global Climate Change, 1998.

Clover, Charles. "Miliband Backs Idea of Carbon Rationing for All." *Daily Telegraph* (London), July 21, 2006.

Coase, Ronald. "The Problem of Social Cost." *Journal of Law and Economics* 3, no. 1 (1960): 1–44.

Cohen, Richard E. *Washington at Work: Back Rooms and Clean Air.* Boston: Allyn and Bacon, 1995.

Cole, Daniel H. *Pollution and Property.* Cambridge: Cambridge University Press, 2002.

Colten, Craig E. *An Unnatural Metropolis: Wresting New Orleans from Nature.* Baton Rouge: Louisiana State University Press, 2005.

Columbia University Master of Arts Program in Climate and Society. "What is Climate and Society?" http://www.columbia.edu/cu/climatesociety/aboutclimate.html.

Columbia University Master of Arts Program in Climate and Society. "From the Director." http://www.columbia.edu/cu/climatesociety/director.html.

Comiso, J. C. "A Rapidly Declining Perennial Sea Ice Cover in the Arctic." *Geophysical Research Letters* 29 (2002): 1956.

Constanza, R. L., L. Wainger, C. Folke, and K. Maler. "Modeling Complex Ecological Economic Systems: Toward an Evolutionary, Dynamic Understanding of People and Nature." *BioScience* 43, no. 8 (1993): 545–555.

Conybeare, John A. C. *Merging Traffic: The Consolidation of the International Automobile Industry.* Lanham, MD: Rowman & Littlefield, 2004.

Covington, William W. "Helping Western Forests Heal." *Nature* 408 (2000): 135–136.

Daily, Gretchen C., and Joshua S. Reichert, eds. *Nature's Services: Societal Dependence on Natural Ecosystems.* Washington, DC: Island Press, 1997.

Daly, Emma. "Helping Students Cope with a Katrina-Tossed World." *New York Times*, November 16, 2005, B9.

Davis, David. *Energy Politics.* New York: St. Martin's Press, 1993.

Dawson, Michael. *The Consumer Trap: Big Business Marketing in American Life.* Chicago: University of Illinois Press, 2003.

Dear, Michael J. *The Postmodern Urban Condition.* Oxford: Blackwell, 2000.

DeBerry, Jarvis. "When Life's Haven Turns Deathtrap." *New Orleans Times-Picayune*, November 6, 2005, 7 (Metro).

Declaration of the United Nations Conference on the Human Environment. 1972. Stockhom: United Nations Environment Programme. Agency.

Deffeyes, Kenneth S. *Hubbert's Peak: The Impending World Oil Shortage.* Princeton, NJ: Princeton University Press, 2001.

Deleuze, Gilles. *What Is Philosophy?* New York: Columbia University Press, 1994.

Dennis, Carina. "Burning Issues." *Nature* 421 (January 2003): 204–206.

DeSombre, Elizabeth. "Global Warming: More Common than Tragic." *Ethics and International Affairs* 18 (2004): 41–46. Dewees, Donald N. "The Decline of the American Street Railways." *Traffic Regulation* 24 (1970): 563–581.

Dewhurst, Frederic, and the Twentieth Century Fund. *America's Needs and Resources.* New York: Twentieth Century Fund, 1955.

Dietz, T., E. Ostrom, and P. C. Stern. "The Struggle to Govern the Commons." *Science* 302 (2003): 1907–1912.

Dimitrov, R. S. "Knowledge, Power and Interests in Environmental Regime Formation." *International Studies Quarterly* 47 (2003): 123–150.

Domack, E., D. Duran, A. Leventer, S. Ishman, S. Doane, S. McCallum, D. Amblas, J. Ring, R. Gilbert, and M. Prentice. "Stability of the Larsen B Ice Shelf on the Antarctic Peninsula During the Holocene Epoch." *Nature* 436, no. 7051 (August 4, 2005): 681–685.

Domeck, M. P., J. E. Williams, and C. A. Wood. "Wildfire Policy and Public Lands: Integrating Scientific Understanding with Social Concerns across Landscapes." *Conservation Biology* 18, no. 4 (2004): 883–889.

Domhoff, G. William. *Who Rules America?* New York: McGraw-Hill, 2005.

Dryzek, John. *The Politics of the Earth.* 2nd ed. New York: Oxford University Press, 2005.

Du Boff, Richard B. *Accumulation & Power: An Economic History of the United States.* Armonk, NY: M. E. Sharpe, 1989.

Dworkin, Ronald. "What Is Equality? Part 1: Equality of Welfare." *Philosophy and Public Affairs* 10, no. 3 (1981): 185–251.

Edwards, David V., and Allesandra Lippucci. *Practicing American Politics.* New York: Worth Publishers, 1998.

Eilperin, Juliet. "Shrinking La. Coastline Contributes to Flooding." *Washington Post*, August 30, 2005, A7.

Ellerman, A. D., B. Buchner, and C. Carraro, eds. *Rights, Rents, and Fairness: Allocation in the European Emissions Trading Scheme.* Cambridge: Cambridge University Press, 2007.

Elliott, Lorraine. *The Global Politics of the Environment.* New York: New York University Press, 1998.

Emanuel, Kerry. "Increasing Destructiveness of Tropical Cyclones Over the Past 30 Years." *Nature* 436, no. 7051 (August 4, 2005): 686–688.

Energy Information Administration (EIA). *Annual Energy Review 2003.* Washington, DC: U.S. Department of Energy, 2004.

Fackler, Martin. "Toyota's Profit Soars, Helped by U.S. Sales." *New York Times*, August 5, 2006.

Fearon, Peter. *War, Prosperity and Depression: The U.S. Economy 1917–45*. Lawrence: University Press of Kansas, 1987.

Field, Alexander J. "Technological Change and U.S. Productivity Growth in the Interwar Years." *Journal of Economic History* 66, no. 1 (2006): 203–236.

Fishman, Robert. *Bourgeois Utopias: The Rise and Fall of Suburbia*. New York: Basic Books, 1987.

Flink, James. *The Car Culture*. Cambridge, MA: MIT Press, 1975.

Flink, James. *The Automobile Age*. Cambridge, MA: MIT Press, 1990.

Fogelson, Robert M. *Bourgeois Nightmares: Suburbia, 1870–1930*. New Haven, CT: Yale University Press, 2005.

Foley, Duncan K. *Understanding Capital: Marx's Economic Theory*. Cambridge, MA: Harvard University Press, 1986.

Foster, Mark S. "The Model-T, the Hard Sell, and Los Angeles's Urban Growth: The Decentralization of Los Angeles during the 1920s." *Pacific Historical Review* 44 (1975): 459–484.

Foster, Mark S. *From Streetcar to Superhighway: American City Planners and Urban Transportation, 1900–1940*. Philadelphia: Temple University Press, 1981.

Fox, Shari. *When the Weather Is Uggianaqtuq: Inuit Observations of Environmental Change*. CD-ROM. Boulder: University of Colorado Geography Department Cartography Lab, 2003. Distributed by National Snow and Ice Data Center.

Fraser, Nancy. *Justice Interruptus: Critical Reflections on the "Postsocialist" Condition*. New York: Routledge, 1997.

French, Hilary F. *Vanishing Borders: Protecting the Planet in the Age of Globalization*. New York: Norton, 2000.

Frey, William H., and Audrey Singer. *Katrina and Rita Impacts on Gulf Coast Populations: First Census Findings*. Washington, DC: Brookings Institution, 2006.

Frey, William H., Audrey Singer, and David Park. *Resettling New Orleans: The First Full Picture from the Census*. Washington, DC: Brookings Institution, 2007.

Frumkin, Norman. *Tracking America's Economy*. 4th ed. Armonk, NY: M. E. Sharpe, 2004.

Gabe, Thomas, Gene Falk, and Maggie McCarty. *Hurricane Katrina: Social-Demographic Characteristics of Impacted Areas*. Washington, DC: Congressional Research Service, 2005.

Galdorisi, George V., and Kevin R. Vienna. *Beyond the Law of the Sea: New Directions for U.S. Oceans Policy*. Westport, CT: Praeger Press, 1997.

Gardiner, Stephen M. "The Real Tragedy of the Commons." *Philosophy and Public Affairs* 30 (2001): 387–416.

Gardiner, Stephen M. "The Pure Intergenerational Problem." *Monist* 86 (2003): 481–500.

Gardiner, Stephen M. "The Global Warming Tragedy and the Dangerous Illusion of the Kyoto Protocol." *Ethics and International Affairs* 18 (2004): 23–39.

Gardiner, Stephen M. "Ethics and Global Climate Change." *Ethics* 114 (2004): 555–600.

Gardiner, Stephen M. "Saved by Disaster? Abrupt Climate Change, Political Inertia, and the Possibility of an Intergenerational Arms Race." *Journal of Social Philosophy*. Forthcoming.

Gardiner, Stephen M. "Why Do Future Generations Need Protection?" Working paper. Paris: Chaire Developpment Durable, 2006. http://ceco.polytechnique.fr/CDD/PDF/DDX-06-16.pdf.

Gonzalez, George A. *Corporate Power and the Environment*. Lanham, MD: Rowman & Littlefield, 2001.

Gonzalez, George A. "Ideas and State Capacity, or Business Dominance? A Historical Analysis of Grazing on the Public Grasslands." *Studies in American Political Development* 15 (2001): 234–244.

Gonzalez, George A. "The Comprehensive Everglades Restoration Plan: Economic or Environmental Sustainability?" *Polity* 37, no. 4 (2005): 466–490.

Gonzalez, George A. *The Politics of Air Pollution*. Albany: State University of New York Press, 2005.

Gonzalez, George A. "Urban Sprawl, Global Warming, and the Limits of Ecological Modernization." *Environmental Politics* 14, no. 3 (2005): 344–362.

Gordon, Robert. "Critical Legal Studies Symposium: Critical Legal Histories." *Stanford Law Review* 36, no. 1/2 (January 1984): 57–125.

Greider, William. *One World, Ready or Not: The Manic Logic of Global Capitalism*. New York: Simon & Schuster, 1997.

Grubb, Michael. *The Greenhouse Effect: Negotiating Targets*. London: Royal Institute of International Affairs, 1989.

Grubb, Michael. *Energy Policies and the Greenhouse Effect*. Aldershot, UK: Gower, 1990.

Grubb, Michael, C. Vrolijk, and D. Brack. *The Kyoto Protocol: A Guide and Assessment*. London: Royal Institute of International Affairs, 1999.

Gunderson, Lance, and C. S. Holling, eds. *Panarchy: Understanding Transformations in Human and Natural Systems*. Washington, DC: Island Press, 2002.

Haas, Peter, Marc Levy, and Ted Parson. "Appraising the Earth Summit: How Should We Judge UNCED's Success?" *Environment* 34, no. 8 (1992): 6–33.

Hampson, Fen Osler, and Judith Reppy, eds. *Earthly Goods: Environmental Change and Social Justice*. Ithaca, NY: Cornell University Press, 1996.

Hanna, Susan, and Svein Jentoft. "Human Use of the Natural Environment: An Overview of Social and Economic Dimensions." In Susan Hanna, Carle Folke, and Karl-Goran Maler, eds., *Rights to Nature: Ecological, Economic, Cultural, and Political Principles of Institutions for the Environment*. Washington, DC: Island Press, 1996.

Hannesson, Rögnvaldur. *The Privatization of the Oceans*. Cambridge, MA: MIT Press, 2004.

Hansen, James, and Makiko Sato. "Greenhouse Gas Growth Rates." *Proceedings of the National Academy of Sciences* 101, no. 46 (2004): 16109–16114.

Hansen, James. "Can We Still Avoid Dangerous Human-Made Climate Change?" Talk presented at the New School University, February 2006.

Haraway, Donna. *Simians, Cyborgs, and Women*. New York: Routledge, 1991.

Hardin, Garrett. "The Tragedy of the Commons." *Science* 162 (1968): 1243–1248.

Harvey, David. *The Condition of Postmodernity*. Oxford: Blackwell, 1989.

Haugland, Torleif, Helge Ole Bergensen, and Kjell Roland. *Energy Structures and Environmental Futures*. New York: Oxford University Press, 1998.

Hayden, Dolores. *Building Suburbia*. New York: Pantheon, 2003.

Hayward, Tim. *Constitutional Environmental Rights*. New York: Oxford University Press, 2005.

Hecht, Jeff. "Six Deadly Deltas, Eight Million People." *New Scientist*, February 18, 2006, 8.

Henderson, Shirley. "After the Hurricanes: Opening Hearts & Homes to the Evacuees." *Ebony*, December 2005, 162.

Henthorn, Cynthia Lee. *From Submarines to Suburbs: Selling a Better America, 1939–1959*. Columbus: Ohio State University Press, 2006.

Herzog, Howard, Balour Eliasson, and Olav Kaarstad. "Capturing Greenhouse Gases." *Scientific American* 282, no. 2 (February 2000): 72–79.

Hirt, Paul W. *A Conspiracy of Optimism: Management of the National Forests since World War Two*. Lincoln: University of Nebraska Press, 1994.

Hise, Greg. *Magnetic Los Angeles: Planning the Twentieth-Century Metropolis*. Baltimore: Johns Hopkins University Press, 1997.

Holling, C. S. "What Barriers? What Bridges?" In L. H. Gunderson, C. S. Holling, and S. S. Light, eds., *Barriers and Bridges to the Renewal of Ecosystems and Institutions*. New York: Columbia University Press, 1995.

Holmes, Bob. "Melting Ice, Global Warning." *New Scientist*, October 2, 2004, 8.

Horne, Jed. *Breach of Faith: Hurricane Katrina and the Near Death of a Great American City*. New York: Random House, 2006.

Hornstein, Jeffrey M. *A Nation of Realtors: A Cultural History of the Twentieth-Century American Middle Class.* Durham, NC: Duke University Press, 2005.

Hughes, Jonathan. *Ecology and Historical Materialism.* New York: Cambridge University Press, 2000.

Hurst, Willmarine. "Longing for Home and How It Used to Be." *New Orleans Times-Picayune,* November 14, 2005, 5 (Metro).

Hyman, Sidney. *Marringer S. Eccles.* Stanford, CA: Stanford University Graduate School of Business, 1976.

Ingham, Richard. "New Orleans Disaster Serves Up a Tough Lesson on Environment." *Agence France-Press,* September 1, 2005. http://www.afp.com/english/globe/?pid=archives.

Intergovernmental Panel on Climate Change (IPCC). *Assessment Report.* Geneva: Intergovernmental Panel on Climate Change, 1990.

Intergovernmental Panel on Climate Change. *Second Assessment Report.* Geneva: Intergovernmental Panel on Climate Change, 1996.

Intergovernmental Panel on Climate Change. *Climate Change 2001: Synthesis Report.* Cambridge: Cambridge University Press, 2002. www.ipcc.ch.

Intergovernmental Panel on Climate Change. Summary for Policymakers. *Climate Change 2001: Impacts, Adaptation, and Vulnerability.* Cambridge: Cambridge University Press, 2001. www.ipcc.ch.

Intergovernmental Panel on Climate Change. *Third Assessment Report.* Geneva: Intergovernmental Panel on Climate Change, 2001.

Intergovernmental Panel on Climate Change. *Climate Change 2007: The Physical Science Basis, Summary for Policymakers.* Geneva: World Meteorological Organization and the United Nations Environment Programme, 2007.

Intergovernmental Panel on Climate Change. *Climate Change 2007: Synthesis Report,* "Summary for Policymakers" (draft report), www.ipcc.ch.

Isser, Steve. *The Economics and Politics of the United States Oil Industry, 1920–1990.* New York: Garland, 1996.

Jackson, Kenneth T. *Crabgrass Frontier: The Suburbanization of the United States.* New York: Oxford University Press, 1985.

Jacob, Klaus. "Time for a Tough Question: Why Rebuild?" *Washington Post,* September 6, 2005, A25.

Jacobson, Linda. "Hurricanes' Aftermath Is Ongoing." *Education Week* 25, no. 21 (2006): 1–18.

James, Rosemary. "New Orleans Is a Pousse-Café." In Rosemary James, ed., *My New Orleans: Ballads to the Big Easy by Her Sons, Daughters, and Lovers,* 1–20. New York: Simon and Schuster, 2006.

Jameson, Fredric. *Postmodernism, or the Cultural Logic of Late Capitalism.* Durham, NC: Duke University Press, 1991.

Jamieson, Dale. "Ethics, Public Policy, and Global Warming." *Science, Technology, & Human Values* 17, no. 2 (1992): 139–153.

Jordan, V. "Sins of Omission." *Environmental Action* 11 (1980): 26–27.

Junger, Sebastian. *The Perfect Storm: A True Story of Men against the Sea*. New York: Norton, 1997.

Kaplan, Robert D. *The Ends of the Earth: A Journey at the Dawn of the 21st Century*. New York: Random House, 1996.

Kelsen, Hans. *Principles of International Law*. New York: Rinehart & Company, 1950.

Kenworthy, Jeffrey R., and Felix B. Laube, with Peter Newman, Paul Barter, Tamim Raad, Chamlong Poboon, and Benedicto Guia Jr. *An International Sourcebook of Automobile Dependence in Cities 1960–1990*. Boulder: University Press of Colorado, 1999.

Kirlin, John. "What Government Must Do Well: Creating Value for Society." *Journal of Public Administration Research and Theory* 6, no. 1 (1996): 161–185.

Kiss, Alexandre., and Dinah Shelton. *International Environmental Law*. New York: Transnational, 1991.

Krasner, Stephen D. "Structural Causes and Regime Consequences: Regimes as Intervening Variables." *International Organizations* 36, no. 21 (1982): 185–205.

Lamborn, Alan C., and Joseph Lepgold. *World Politics into the 21st Century*. Preliminary ed. Upper Saddle River, NJ: Prentice Hall, 2003.

Lanne, Markku, and Matti Liski. "Trends and Breaks in Per-Capita Carbon Dioxide Emissions, 1870–2028." *Energy Journal* 25, no. 4 (2004): 41–65.

Laswell, Harold. *Politics: Who Gets What, When, and How*. New York: McGraw-Hill, 1936.

Latour, Bruno. *The Politics of Nature*. Cambridge, MA: Harvard University Press, 2004.

Lebel, L., N. H. Tri, A. Saengnoree, S. Pasong, U. Buatama, and L. K. Thoa. "Industrial Transformation and Shrimp Aquaculture in Thailand and Vietnam: Pathways to Ecological, Social, and Economic Sustainability?" *Ambio* 31, no. 4 (2002): 311–322.

Lepert, Beate. "Observed Reductions in Surface Solar Radiation at Sites in the U.S. and Worldwide." *Geophysical Research Letters* 29, no. 10 (2002): 1421–1433.

Lester, James P., David W. Allen, and Kelly M. Hill. *Environmental Injustice in the United States: Myths and Realities*. Boulder, CO: Westview Press, 2001.

Lewis, Peirce F. *New Orleans: The Making of an Urban Landscape*. Santa Fe, NM: Center for American Places, 2003.

Lindblom, Charles E. *Politics and Markets: The World's Political-Economic Systems*. New York: Basic Books, 1977.

Liu, Amy, and Allison Plyer. *New Orleans Index, Second Anniversary Edition: A Review of Key Indicators of Recovery Two Years After Katrina*. Washington,

DC: Brookings Institution and Greater New Orleans Community Data Center, 2007.

Logan, John, and Harvey Molotch. *Urban Fortunes: The Political Economy of Place*. Berkeley: University of California Press, 1988.

Lomborg, Bjørn. *The Skeptical Environmentalist: Measuring the Real State of the World*. Cambridge: Cambridge University Press, 2001.

Long, Douglas. *Global Warming*. New York: Facts on File, 2004.

Long, Rob. "Different Down There: In America, But Not Entirely." *National Review*, September 26, 2005, 32, 34.

Longman, Jere. "From the Ruins of Hurricane Katrina Come a Championship and Concerns." *New York Times*, January 22, 2006, 1 (sec. 8).

"The Long Slog After Katrina." Editorial. *Boston Globe*, August 29, 2007, A16.

Lopez, Barry. *About This Life: Journeys on the Threshold of Memory*. New York: Knopf, 1998.

Lovecraft, A. L., and S. F. Trainor. "Organizational Learning and Policy Change in Wildland Fire Agencies: Cases of Uncharacteristic Wildfires in Alaska and Yukon Territory." Unpublished ms., 2006.

Low, Nicholas, and Brendan Gleeson. *Justice, Society and Nature: An Exploration of Political Ecology*. London: Routledge, 1998.

Luke, Timothy W. "Placing Powers, Siting Spaces: The Politics of Global and Local in the New World Order." *Environment and Planning A: Society and Space* 12 (1994): 613–628.

Luke, Timothy W. "Liberal Society and Cyborg Subjectivity: The Politics of Environments, Bodies, and Nature." *Alternatives: A Journal of World Policy* XXI, no. 1 (1996): 1–30.

Luke, Timothy W. "At the End of Nature: Cyborgs, Humachines, and Environments in Postmodernity." *Environment and Planning A*, 29 (1997): 1367–1380.

Luke, Timothy W. "Environmentality as Green Governmentality." In Eric Darier, ed., *Discourses of the Environment*. Oxford: Blackwell, 1999.

Luke, Timothy W. "Training Eco-Managerialists: Academic Environmental Studies as a Power/Knowledge Formation." In Frank Fischer and Maarten Hajer, eds., *Living with Nature: Environmental Discourse as Cultural Politics*. Oxford: Oxford University Press, 1999.

Luke, Timothy W. "Cyborg Enchantments: Commodity Fetishism and Human/Machine Interactions." *Strategies* 13, no. 1 (2000): 39–62.

Luke, Timothy W. "Reconstructing Nature: How the New Informatics Are Rewriting Place, Power, and Property as Bitspace." *Capitalism, Nature, Socialism* 12, no. 3 (September 2001): 3–27.

Luke, Timothy W. "Global Cities vs. global cities: Rethinking Contemporary Urbanism as Public Ecology." *Studies in Political*. Economy 71 (2003): 11–22.

Luke, Timothy W. "The System of Sustainable Degradation." *Capitalism, Nature, Socialism* 17, no. 1 (2006): 99–112.

Luper-Foy, Steven. "Justice and Natural Resources." *Environmental Values* 1, no. 1 (spring 1992): 47–64.

Lyotard, Jean-François. *The Postmodern Condition: A Report on Knowledge.* Minneapolis: University of Minnesota Press, 1984.

Manley, John F. "Neo-Pluralism." *American Political Science Review* 77, no. 2 (1983): 368–383.

Marland, G., T. Boden, and R. J. Andreas. "Global CO_2 Emissions from Fossil-Fuel Burning, Cement Manufacture, and Gas Flaring: 1751–2002." Carbon Dioxide Information Analysis Center, U.S. Department of Energy. 2005. http://cdiac.ornl.gov/trends/emis/glo.htm.

Marshall, Bob. "The River Wild: Rebuilding Our Coastal Wetlands the Old-Fashioned Way." *New Orleans Times-Picayune*, March 10, 2006, 7.

Marx, Karl. *Grundrisse.* New York: Vintage, 1973.

Maslin, Mark. *Global Warming.* Oxford: Oxford University Press, 2004.

McCaffrey, Sarah. "Thinking of Wildfire as a Natural Hazard." *Society and Natural Resources* 17 (2004): 509–516.

McCarthy, Kevin, D. J. Peterson, Narayan Sastry, and Michael Pollard. *The Repopulation of New Orleans After Hurricane Katrina.* Santa Monica, CA: RAND Corporation, 2006.

McFarland, Andrew S. *Neopluralism.* Lawrence: University Press of Kansas, 2004.

McNeill, J. R. *Something New under the Sun: An Environmental History of the Twentieth-Century World.* New York: Norton, 2000.

McPhee, John. *The Control of Nature.* New York: Farrar, Straus and Giroux, 1990.

Meehl, Gerald, Warren M. Washington, William D. Collins, Julie M. Arblaster, Aixue Hu, Lawrence E. Buja, Warren G. Strand, and Haiyan Teng. "How Much More Global Warming and Sea Level Rise?" *Science* 307 (2005): 1769–1772.

Mendelsohn, Robert O. *Global Warming and the American Economy.* London: Edward Elgar, 2001.

Meyer, Aubrey. "The Kyoto Protocol and the Emergence of 'Contraction and Convergence' as a Framework for an International Political Solution to Greenhouse Gas Emissions Abatement." In Olav Hohmayer and Klaus Rennings, eds., *Man-Made Climate Change: Economic Aspects and Policy Options.* Mannheim: Zentrum für Europäische Wirtschaftsforschung (ZEW), 1999.

Meyer, John M. *Political Nature: Environmentalism and the Interpretation of Western Thought.* Cambridge, MA: MIT Press, 2001.

Mielke, James E. *Deep Seabed Mining: U.S. Interests and the U.N. Convention on the Law of the Sea.* Washington, DC: Congressional Research Service, 1995.

Mielke, James E. *Polar Research: U.S. Policy and Interests*. Washington, DC: Congressional Research Service, 1996.

Milenky, E. S., and S. I. Schwab. "Latin America and Antarctica." *Current History* 82 (1983): 52.

Miles, Edward L. *Global Ocean Politics: The Decision Process at the Third United Nations Conference on the Law of the Sea 1973–1982*. The Hague: Martinus Nijhoff, 1998.

Miliband, Ralph. *The State in Capitalist Society*. New York: Basic Books, 1969.

Millennium Ecosystem Assessment. *Ecosystems and Human Well-Being: Synthesis*. Washington, DC: Island Press, 2005.

Miller, Edward. "Some Implications of Land Ownership Patterns for Petroleum Policy." *Land Economics* 49, no. 4 (1973): 414–423.

Mohai, Paul, and Bunyan Bryant. "Environmental Racism: Reviewing the Evidence." In B. Bryant and P. Mohai, eds., *Race and the Incidence of Environmental Hazards: A Time for Discourse*. Boulder, CO: Westview Press, 1992.

Mooney, Chris. "Warmed Over." *The American Prospect*, online ed., January 10, 2005.

Muller, Peter. *Contemporary Suburban America*. Englewood Cliffs, NJ: Prentice-Hall, 1981.

Mulrine, Anna. "The Long Road Back." *U.S. News & World Report*, February 27, 2006, 44–50, 52–58.

Mumford, Lewis. *The City in History: Its Origins, Its Transformations, and Its Prospects*. New York: Harcourt, Brace & World, 1961.

Natcher, David C. "Implications of Fire Policy on Native Land Use in the Yukon Flats, Alaska." *Human Ecology* 32, no. 4 (2004): 421–441.

National Academy of Sciences/National Research Council. *Changing Climate*. Washington, DC: National Academy Press, 1983.

National Assessment Synthesis Team. *Climate Change Impacts on the United States: The Potential Consequences of Climate Variability and Change*. Cambridge: Cambridge University Press, 2000. www.usgcrp.gov/usgcrp/nacc/default.htm

National Center for Children in Poverty, Columbia University. "Child Poverty in States Hit by Hurricane Katrina." Fact Sheet No. 1. September 2005. www.nccp.org/pub_cpt05a.html.

Newman, Peter, and Jeffrey Kenworthy. *Sustainability and Cities: Overcoming Automobile Dependence*. Washington, DC: Island Press, 1999.

Nitze, William A. "A Failure of Presidential Leadership." In Irving Mintzer and J. Amber Leonard, eds., *Negotiating Climate Change: The Inside Story of the Rio Convention*. Cambridge: Cambridge University Press, 1994.

NOAA. "Changes in Arctic Sea Ice over the Past 50 Years: Bridging the Knowledge Gap between Scientific Community and Alaska Native Community."

Executive Summary from the Marine Mammal Commission Workshop on the Impacts of Changes in Sea Ice and Other Environmental Parameters in the Arctic. 2000. http://www.arctic.noaa.gov/workshop_summary.html.

NOAA. Arctic Change: A Near-Realtime Arctic Change Indicator. 2006 (updated November). http://www.arctic.noaa.gov/detect/human-shishmaref.shtml.

O'Connor, James. "Capitalism, Nature, Socialism: A Theoretical Introduction." *Capitalism, Nature, Socialism* 1 (fall 1988): 11–39.

Odum, Howard T. *Systems Ecology.* New York: Wiley, 1983.

Odum, W. E., E. P. Odum, and H. T. Odum. "Nature's Pulsing Paradigm." *Estuaries* 18, no. 4 (1995): 547–555.

Ohmura, Atsumu. "Reevaluation and Monitoring of the Global Energy Balance." In M. Sanderson, ed., *UNESCO Source Book in Climatology.* Paris: UNESCO, 1990.

Olegario, Rowena. *A Culture of Credit: Embedding Trust and Transparency in American Business.* Cambridge, MA: Harvard University Press, 2006.

Olney, Martha L. "Credit as a Production-Smoothing Device: The Case of Automobiles, 1913–1938." *Journal of Economic History* 27, no. 2 (1989): 322–349.

Olney, Martha L. *Buy Now, Pay Later: Advertising, Credit, and Consumer Durables in the 1920s.* Chapel Hill: University of North Carolina Press, 1991.

Olson, Mancur. *The Logic of Collective Action.* Cambridge, MA: Harvard University Press, 1971.

O'Neill, Brian C., and Michael Oppenheimer. "Dangerous Climate Impacts and the Kyoto Protocol." *Science* 296 (2002): 1971–1972.

Ostrom, Elinor. *Governing the Commons: The Evolution of Institutions for Collective Action.* Cambridge: Cambridge University Press, 1990.

O'Toole, Randal. *Reforming the Forest Service.* Washington, DC: Island Press, 1988.

Ott, Hermann E., and Wolfgang Sachs. *Ethical Aspects of Emissions Trading.* Wuppertal, Germany: Wuppertal Institute, 2000.

Pacala, Stephen., and Robert Socolow. "Stabilization Wedges: Solving the Climate Problem for the Next 50 Years with Current Technologies." *Science* 305 (August 13, 2004): 968–972.

Pan, Philip P. "Scientists Issue Dire Prediction on Warming." *Washington Post.* January 23, 2001, A1.

Parfit, Michael. "The Last Continent." *Smithsonian* 15 (1984): 50–60.

Parks, Bradley C., and J. Timmons Roberts. "Environmental and Ecological Justice." In Michele M. Betsill, Kathryn Hochstetler, and Dimitris Stevis, eds., *Palgrave Advances in International Environmental Politics.* Basingstoke, UK: Palgrave Macmillan, 2006.

Parra, Francisco. *Oil Politics.* New York: I. B. Tauris, 2005.

Parsons, Howard L. "Introduction." In Howard L. Parsons, ed. and comp., *Marx and Engels on Ecology*. Westport, CT: Greenwood Press, 1977.

Pauly, Robert J. *U.S. Foreign Policy and the Persian Gulf*. Burlington, VT: Ashgate, 2005.

Pearce, Fred. "The Flaw in the Thaw." *New Scientist*, August 27, 2005, 26.

Perelman, Michael. *The Perverse Economy*. New York: Palgrave Macmillan, 2003.

Philip, George. *The Political Economy of International Oil*. Edinburgh: Edinburgh University Press, 1994.

Piazza, Tom. *Why New Orleans Matters*. New York: HarperCollins, 2005.

Plumwood, Val. "Nature, Self, and Gender: Feminism, Environmental Philosophy, and the Critique of Rationalism." In Michael Zimmerman, J. Baird Callicott, George Sessions, Karen J. Warren, and John Clark, eds., *Environmental Philosophy: From Animal Rights to Radical Ecology*. Englewood Cliffs, NJ: Prentice Hall, 1993.

"Politics May Doom New Orleans." Editorial. *San Francisco Chronicle*, August 25, 2006, B10.

Porter, G., and J. W. Brown. *Global Environmental Politics*. Boulder, CO: Westview Press, 1996.

Pritzsche, Kai. "Development of the Concept of Common Heritage of Mankind Outer Space Law and Its Contents in the 1979 Moon Treaty." Master Dissertation. Berkeley: School of Law, University of California at Berkeley, 1984.

Proudhon, P.-J. *What Is Property?* Cambridge: Cambridge University Press, [1840] 1993.

Putnam, Robert. *Making Democracy Work*. Princeton, NJ: Princeton University Press, 1993.

Pyne, Stephen J. *Fire in America: A Cultural History of Wildland and Rural Fire*. Princeton, NJ: Princeton University Press, 1982.

Rawls, John. *A Theory of Justice*. Cambridge, MA: Belknap Press of Harvard University Press, 1971.

Raymond, Leigh. *Private Rights in Public Resources: Equity and Property Allocation in Market-Based Environmental Policy*. Washington, DC: Resources for the Future Press, 2003.

Raymond, Leigh. "Allocating Greenhouse Gas Emissions Under the EU ETS: The UK Experience." Paper presented at the Sixth Open Meeting of the Human Dimensions of Global Environmental Change Research Community, University of Bonn, Germany, 2005.

Raymond, Leigh. "Viewpoint: Cutting the 'Gordian Knot' in Climate Change Policy." *Energy Policy* 34 (2006): 655–658.

Real Climate. "Global Dimming." January 18, 2005. http://www.realclimate.org/index.dhd?p=los.

Reice, Seth. *The Silver Lining: The Benefits of Natural Disasters.* Princeton, NJ: Princeton University Press, 2001.

Repetto, Robert. "The Clean Development Mechanism: Institutional Breakthrough or Institutional Nightmare?" *Policy Sciences* 34 (2001): 303–327.

Ricardo, David. *On the Principles of Political Economy and Taxation.* Washington, DC: J. B. Bell, 1830.

Rivlin, Gary. "Wealthy Blacks Oppose Plans for Their Property." *New York Times,* December 10, 2005, A11.

Robbins, William G. *Lumberjacks and Legislators: Political Economy of the U.S. Lumber Industry, 1890–1941.* College Station: Texas A&M University Press, 1982.

Roberts, Paul. *The End of Oil.* New York: Houghton Mifflin, 2004.

Roderick, Michael, and Gerald Farquhar. "The Cause of Decreased Pan Evaporation over the Past 50 Years." *Science* 298 (2002): 1410–1411.

Rohter, Larry. "Antarctica, Warming, Looks Ever More Vulnerable." *New York Times,* January 25, 2005, F1.

Roig-Franzia, Manuel. "A City Fears for Its Soul: New Orleans Worries That Its Unique Culture May Be Lost." *Washington Post,* February 3, 2006, A1.

Rose, Adam, and Brandt Stevens. "The Efficiency and Equity of Marketable Permits for CO_2 Emissions." *Resource and Energy Economics* 15 (1993): 117–146.

Rosen, Elliot. *Roosevelt, the Great Depression, and the Economics of Recovery.* Charlottesville: University of Virginia Press, 2005.

Rosenberg, Merri. "Displaced by Katrina, Coping in a New School." *New York Times,* December 18, 2005, 2 (Westchester Weekly Section).

Rosenthal, Elizabeth. "UN Report Describes Risks of Inaction on Climate Change." *New York Times,* November 17, 2007, A1.

Rotstayn, Leon D., and Ulrike Lothmann. "Observed Reductions in surface Solar Radiation at Sites in the U.S. and Worldwide." *Geophysical Research Letters* 29, no. 10 (2002): 1421–1433.

Rutledge, Ian. *Addicted to Oil: America's Relentless Drive for Energy Security.* New York: I. B. Tauris, 2005.

Sack, Robert David. *Homo Geographicus: A Framework for Action, Awareness, and Moral Concern.* Baltimore: Johns Hopkins University Press, 1997.

Sagar, Ambuj. "Wealth, Responsibility, and Equity: Exploring an Allocation Framework for Global GHG Emissions." *Climatic Change* 45 (2000): 511–527.

Sagoff, Mark. *The Economy of the Earth.* Cambridge: Cambridge University Press, 1988.

Sagoff, Mark. "Settling America: The Concept of Place in Environmental Politics." In Philip D. Brick and R. McGreggor Cawley, eds., *A Wolf in the Garden: The Land Rights Movement and the New Environmental Debate.* Lanham, MD: Rowman & Littlefield, 1996.

Sale, Kirkpatrick. *Dwellers in the Land: The Bioregional Vision.* San Francisco: Sierra Club Books, 1985.

Sampson, R. Neil. "Primed for a Firestorm." *Forum for Applied Research and Public Policy* 14, no. 1 (1999): 20–25.

Samuelson, Robert J. "Lots of Gain and No Pain!" *Newsweek*, February 21, 2005.

Sanders, M. Elizabeth. *The Regulation of Natural Gas.* Philadelphia: Temple University Press, 1981.

Saulny, Susan. "A Legacy of the Storm: Depression and Suicide." *New York Times.* June 21, 2006, A1.

Schlosberg, David. "Reconceiving Environmental Justice: Global Movements and Political Theories." *Environmental Politics* 13, no. 3 (autumn 2004): 517–540.

Schrogl, Kai-Uwe. "Legal Aspects Related to the Application of the Principle That the Exploration and Utilization of Outer Space Should Be Carried Out for the Benefits and in the Interest of All States Taking into Account the Needs of Developing Countries." In M. Benko and K.-U. Schrogl, eds., *International Space Law in the Making.* Gif-Sur-Yvette Cedex. Editions Frontières, 1993.

Schulte, Bret. "Turf Wars in the Delta." *U.S. News & World Report* 140, no. 7 (February 27, 2006): 66–71.

Sen, Amartya. *Inequality Reexamined.* Cambridge, MA: Harvard University Press, 1992.

Shaffer, Ed. *The United States and the Control of World Oil.* New York: St. Martin's Press, 1983.

Shaw, Anup. "Global Dimming." *Global Issues.* Weblog. January 15, 2005. www.globalissues.org/EnvIssues/GlobalWarming/Globaldimming.asp.

Shepski, Lee. "Prisoner's Dilemma: The Hard Problem." Paper presented at the meeting of the Pacific Division of the American Philosophical Association, March 2006.

Shue, Henry. *Basic Rights.* Princeton, NJ: Princeton University Press, 1980.

Shue, Henry. "Subsistence Emissions and Luxury Emissions." *Law and Policy* 15, no. 1 (1993): 39–59.

Shue, Henry. "Global Environment and International Inequality." *International Affairs* 75 (1999): 531–545.

Shue, Henry. "Climate." In Dale Jamieson, ed., *A Companion to Environmental Philosophy.* Malden, MA: Blackwell, 2001.

Shue, Henry. "Responsibility of Future Generations and the Technological Transition." In Walter Sinnott-Armstrong and Richard Howarth, eds., *Perspectives on Climate Change: Science, Economics, Politics, Ethics.* New York: Elsevier, 2005.

Singer, Peter. *One World: The Ethics of Globalization.* New Haven, CT: Yale University Press, 2002.

Smil, Vaclav. *Energy in World History.* Boulder, CO: Westview Press, 1994.

Smith, K. R. "The Natural Debt: North and South." In T. W. Giambelluca and A. Henderson-Sellers, eds., *Climate Change: Developing Southern Hemisphere Perspectives.* Chichester, UK: Wiley, 1996.

Smith, Michael Peter. *Transnational Urbanism: Locating Globalization.* Oxford: Blackwell, 2000.

Socolow, Robert H. "Can We Bury Global Warming?" *Scientific American* 293, no. 1 (July 2005): 49–55.

Soja, Edward. *Postmetropolis: Critical Studies of Cities.* Oxford: Blackwell, 2000.

Sparks, Richard E. "Rethinking, Then Rebuilding New Orleans." *Issues in Science & Technology* 22, no. 2 (winter 2006): 33–39.

Spash, Clive L. *Greenhouse Economics: Value and Ethics.* London: Routledge, 2002.

Stanhill, Gerald, and Shabtai Cohen. "Global Dimming: A Review of the Evidence." *Agricultural and Forest Meteorology* 107 (2001): 255–278.

St. Clair, David J. *The Motorization of American Cities.* New York: Praeger, 1986.

Stevens, William K. *The Change in the Weather: People, Weather, and the Science of Climate.* New York: Delta, 1999.

Stroeve, J. C., M. C. Serreze, F. Fetterer, T. Arbetter, W. Meier, J. Maslanik, and K. Knowles. "Tracking the Arctic's Shrinking Ice Cover: Another Extreme September Minimum in 2004." *Geophysical Research Letters* 32 (2005): L04501. doi:04510.01029/02004GL021810.

Szasz, Andrew. *EcoPopulism: Toxic Waste and the Movement for Environmental Justice.* Minneapolis: University of Minnesota Press, 1994.

Taylor, Prue. *An Ecological Approach to International Law: Responding to Challenges of Climate Change.* London: Routledge, 1998.

Thomas, Robert Paul. *An Analysis of the Pattern of Growth of the Automobile Industry, 1895–1929.* New York: Arno, 1977.

Tokar, Brian. *Earth for Sale: Reclaiming Ecology in the Age of Corporate Greenwash.* Boston: South End Press, 1997.

Traxler, Martino. "Fair Chore Division for Climate Change." *Social Theory and Practice* 28 (2002): 101–134.

Tuan, Yi-Fu. *Space and Place: The Perspective of Experience.* Minneapolis: University of Minnesota Press, 1977.

Tuan, Yi-Fu. *Topophilia: A Study of Environmental Perception, Attitudes, and Values.* New York: Columbia University Press, 1990.

Turner, B. L., II, R. E. Kasperson, P. A. Matson, J. J. McCarthy, R. W. Corell, L. Christensen, N. Eckley, J. X. Kasperson, A. Luers, M. L. Martello, C. Polsky, A. Pulsipher, and A. Schiller. *Proceedings of the National Academy of Sciences, USA* 100 (2003): 8074–8079.

Twentieth Century Fund Task Force on the International Oil Crisis. *Paying for Energy*. Report. New York: McGraw-Hill, 1975.

Twentieth Century Fund Task Force on United States Energy Policy. *Providing for Energy*. Report. New York: McGraw-Hill, 1977.

Tyler, Tom. *Why People Obey the Law*. New Haven, CT: Yale University Press, 1990.

Unger, Roberto Mangabeira. *The Critical Legal Studies Movement*. Cambridge, MA: Harvard University Press, 1986.

United Church of Christ, Commission for Racial Justice. *Toxic Wastes and Race: A National Report on the Racial and Socio-Economic Characteristics of Communities with Hazardous Waste Sites*. New York: United Church of Christ, 1987.

United Nations. *United Nations Framework Convention on Climate Change*. New York: United Nations, 1992.

United Nations University, Institute for Environment and Human Security. "As Ranks of 'Environmental Refugees' Swell Worldwide, Calls Grow for Better Definition, Recognition, Support." Press release. October 12, 2005. http://www.ehs.unu.edu.

U.S. Department of Commerce / National Oceanic and Atmospheric Administration. "NOAA Reviews Record-Setting 2005 Atlantic Hurricane Season." www.noaanews.noaa.gov/stories2005/s2540.htm.

U.S. Department of the Interior. *Alaska Consolidated Interagency Fire Management Plan*. Operational draft. Fairbanks, AK: U.S. Bureau of Land Management, 1998.

Useem, Michael. *The Inner Circle: Large Corporations and the Rise of Business Political Activity in the U.S. and U.K.* Oxford: Oxford University Press, 1984.

U.S. Senate Committee on Commerce, Science, and Transportation. *Report on Agreement Governing the Activities of States on the Moon and Other Celestial Bodies*. Washington, DC: Government Printing Office, 1980.

Vale, Thomas R., ed. *Fire, Native Peoples, and the Natural Landscape*. Washington, DC: Island Press, 2002.

Vanderheiden, Steve. "Knowledge, Uncertainty, and Responsibility: Responding to Climate Change." *Public Affairs Quarterly* 18, no. 2 (2004): 141–158.

Vanderheiden, Steve. *Atmospheric Justice: A Political Theory of Climate Change*. New York: Oxford University Press, 2008.

Van Til, Jon. *Living with Energy Shortfall*. Boulder, CO: Westview Press, 1982.

Victor, David G. *The Collapse of the Kyoto Protocol and the Struggle to Slow Global Warming*. Princeton, NJ: Princeton University Press, 2001.

Victor, David G. *Climate Change: Debating America's Policy Options*. New York: Council on Foreign Relations, 2004.

Vietor, Richard H. *Environmental Politics & the Coal Coalition.* College Station: Texas A&M University Press, 1980.

Vig, Norman J., and Regina S. Axelrod, eds. *The Global Environment: Institutions, Law, and Policy.* Washington, DC: CQ Press, 1999.

Virilio, Paul. *The Art of the Motor.* Minneapolis: University of Minnesota Press, 1995.

Virilio, Paul. *Open Sky.* London: Verso, 1997.

Waldron, Jeremy. "The Advantages and Difficulties of the Humean Theory of Property." *Social Philosophy and Policy* 11 (1994): 85–123.

Waltz, Kenneth. *Theory of International Politics.* New York: McGraw-Hill, 1979.

Wardell, D. A., T. T. Nielsen, K. Rasmussen, and C. Mbow. "Fire History, Fire Regimes and Fire Management in West Africa: An Overview." In J. G. Goldammer and C. de Ronde, eds., *Wildland Fire Management Handbook for Sub-Sahara Africa.* Freiburg: Oneworldbooks, 2004.

Watt-Cloutier, Sheila. "Inuit Circumpolar Conference Testimony." U.S. Senate Committee on Commerce, Science, and Transportation, Washington DC, September 15, 2004. http://www.ciel.org/Publications/McCainHearingSpeech-15Sept04.pdf.

Weiss, Edith Brown. "The Emerging Structure of International Environmental Law." In Norman J. Vig and Regina S. Axelrod, eds., *The Global Environment: Institutions, Law, and Policy.* Washington, DC: CQ Press, 1999.

Weiss, Marc. *The Rise of the Community Builders: The American Real Estate Industry and Urban Land Planning.* New York: Columbia University Press, 1987.

Wenz, Peter. *Environmental Justice.* Albany: State University of New York Press, 1988.

Wenzel, George W. "Warming the Arctic: Environmentalism and Canadian Arctic." In D. Peterson and D. Johnson, eds., *Human Ecology and Climate Change,* 169–184. Bristol, PA: Taylor & Francis, 1995.

Wetherald, Richard T., Ronald J. Stouffer, and Keith W. Dixon. "Committed Warming and Its Implications for Climate Change." *Geophysical Research Letters* 28, no. 8 (2001): 1535–1538.

Whelan, Robert J. *The Ecology of Fire.* New York: Cambridge University Press, 1995.

Wigley, T. M. L. "The Climate Change Commitment." *Science* 307 (2005): 1766–1769.

Wild, M., H. Gilgen, A. Roesch, A. Ohmura, C. Long, E. Dutton, B. Forgan, A. Kallis, V. Russak, and A. Tsvetkov. "From Dimming to Brightening: Decadal Changes in Solar Radiation at Earth's Surface." *Science* 308 (May 6, 2005): 847–848.

Winks, Robin W. *Laurence S. Rockefeller.* Washington, DC: Island Press, 1997.

Wolch, Jennifer, Manuel Pastor Jr., and Peter Drier, eds. *Up against the Sprawl.* Minneapolis: University of Minnesota Press, 2004.

Wolff, Robert Paul. *Understanding Marx: A Reconstruction and Critique of Capital.* Princeton, NJ: Princeton University Press, 1984.

World Commission on Environment and Development. *Our Common Future.* New York: Oxford University Press, 1987.

World Resources Institutes. *World Resources 2002–2004: Decisions for the Earth: Balance, Voice, and Power.* Washington, DC: World Resources Institute, 2003.

Worster, Donald. *Nature's Economy: The Roots of Ecology.* Garden City, NY: Anchor Books, 1979.

Yamin, Farhana, and Joanna Depledge. *The International Climate Change Regime: A Guide to Rules, Institutions and Procedures.* Cambridge: Cambridge University Press, 2004.

Yergin, Daniel. *The Prize: The Epic Quest for Oil, Money, and Power.* New York: Simon & Schuster, 1991.

Yetiv, Steve A. *Awakenings: Global Oil Security and American Foreign Policy.* Ithaca, NY: Cornell University Press, 2004.

Yokota, Yozo. "International Justice and the Global Environment." *Journal of International Affairs* 52, no. 2 (spring 1999): 583–598.

Young, Iris Marion. *Justice and the Politics of Difference.* Princeton, NJ: Princeton University Press, 1990.

Young, Iris Marion. *Intersecting Voices: Dilemmas of Gender, Political Philosophy, and Policy.* Princeton, NJ: Princeton University Press, 1997.

Young, Oran R. *The Institutional Dimensions of Environmental Change: Fit, Interplay, and Scale.* Cambridge, MA: MIT Press, 2002.

Zaun, Todd. "Honda Tries to Spruce Up a Stodgy Image." *New York Times,* March 19, 2005.

Index

Agency, fragmentation of, 27, 30
Agrawal, Arun, 94–96
Alaska, 54, 100, 104–111, 137
Albedo, 128
Antarctic Treaty. *See* Treaties
Arctic ecosystems, 54, 91–92, 115, 125
Arctic Climate Impact Assessment, 91
Arendt, Hannah, 190
Asia-Pacific Partnership on Clean Development and Climate, xi
Atmospheric absorptive capacity, as resource, 13, 46, 55

Barry, Brian, 50
Beckerman, Wilfred, xxii, 177
Beitz, Charles, 60–61
Benchmarks, as regulatory tool, 6–7, 56–57
Berlin Mandate, 13, 57
Biodiversity. *See* Ecosystem resilience
Brundtland Report, 63
Bush, George W., 43, 58, 71, 177
Byrd-Hagel Resolution, xiii, 43

Carbon dioxide. *See* Greenhouse gases
Carbon sequestration, 19–20
Center for Science and Environment, 44–45
China, 44–45, 47, 56–57, 77
Climate change
  adaptation, 101–103, 177–178, 194–195
  and capitalism, 124, 129–130, 136, 138–142, 154–155
  ecosystem services, impact upon, 98–99
  environmental consequences, 31–32, 92–93, 110–111, 121–124
  flooding (*see* New Orleans)
  and future generations, 33–35, 47
  global cooling, xx, 124–125
  global dimming, xx, 126–127, 131
  global inequality of effects, viii, xiii, 30, 44–45, 58, 71–75, 127
  sea ice, effects upon, 110–111
  social consequences, 92–93, 115, 188–189
  urbanization as cause, 130–131, 153, 157–160
Climatology, 121–122, 124, 133, 137–139, 142–144
Coase, Ronald, 18
Collective action, 27–30, 33
Columbia University, 146–148
Common Heritage of Mankind principle, xvii, 11–12, 17–18
Consumerism, 147–148, 157–161
Contraction and convergence, 57–58
Critical Legal Studies, xix, 67, 75–77, 81–82

Democratic legitimacy, 69–70, 80
Differentiated responsibilities, 31, 43–44, 47, 56, 81

Distributive justice 3, 25, 59, 73–74, 83. *See also* Environmental justice
cosmopolitan, 60–62, 64–65, 83
versus recognition, 74, 76–77

Earth Summit. *See* UN Conference on Environment and Development
Ecological resilience, 101–103, 105–106
and biodiversity, 106, 132
robustness, as measure of, 113–115
and vulnerability, 102
Eco-Marxism. *See* Marxism
Ecosystem. *See* Social-ecological systems analysis
Ecosystem service. *See* Climate change
Einstein, Albert, x
Emissions trading, as regulatory tool, 15–16, 18–20, 80
Environmental justice, 68, 71–74
Environmentality, concept of, 94–95
Environmental law
adjudication, 79–82
and climate policy (*see* Kyoto Protocol)
and fairness, 73
indeterminacy of, 82–83
international, 68–70, 75
Environmental political theory, viii–ix, xvi
Environmental subject, concept of, 99–101, 103
Equality
of emissions, 45, 47–48
equal burdens, 5–6
equal efficiency, 6–7
equal rights, 7, 18, 56–57
equal subsistence rights, 7–8, 53, 59–60
Equity, in climate policy, 4, 44–46
European Union (EU), 13, 15–16, 19, 56
Exchange value, 153–155

Federal Emergency Management Agency (FEMA), 177–178

Federal Housing Authority, xxi, 158–160, 166–167

Global commons, 3, 8–13
Global cooling. *See* Climate change
Global dimming. *See* Climate change
Global warming. *See* Climate change
Gordian knot, 4
Gore, Albert Jr., vii
Greenhouse gases, 29, 31, 34, 45–46, 122–124, 126, 127–129. *See also* Climate change

Habermas, Jürgen, 73, 92
Hardin, Garrett, 27–28
Hayward, Tim, 48, 50
Hobbes, Thomas, 3
Home, concept of, 188–189, 191. *See also* Place
Hurricane Katrina. *See* New Orleans

India, 44–45, 56–57
Indeterminacy of law. *See* Environmental law
Industrial Revolution, 132–133, 155
Institutional inadequacy, 28–29, 32
Inter-American Commission on Human Rights, 54
Intergenerational ethics, 25, 33–35
Intergovernmental Panel on Climate Change
formation and role, vii, 123–124, 141–142
scientific predictions, xi, 31, 71, 126, 132, 187
Inuit, 54, 101, 110–111

Justice. *See* Distributive justice

Kant, Immanuel, 45, 73
Ki-Moon, Ban, xi
Kymlicka, Will, 76
Kyoto Protocol
and fairness, 44, 71, 78–79
negotiations toward, 7, 70
policy requirements, 47, 56–57, 64–65, 123, 127

taking legal effect, xii, 4, 70
U.S. withdrawal from, xi, 43, 71

Legal systems. *See* Environmental law
Liability, 45–46
Lifeworld, concept of, xx, 92,
  103–104
Locke, John, 5–6, 16–17, 60
Lyotard, Jean–François, 129–130

Market-based instruments, 4, 155
Marxism, xxi, 138–144, 153–166.
  See also Exchange value
Millennium Ecosystem Assessment,
  98–99
Mississippi Delta, 186, 192, 194
Moral corruption, 36–37
Moral responsibility. *See*
  Responsibility, principle of

New Orleans
  Bring New Orleans Back Coalition
    (BNOBC), 184–185
  ecological vulnerability, 185–187,
    191–194
  Hurricane Katrina, xxii, 177–181
  post-Katrina problems, 179–183
  unique history and culture, 183–184

O'Connor, James, 138–142
Organization of Petroleum Exporting
  Countries (OPEC), 163–164, 166
Outer Continental Shelf leasing, 111

Perfect storm, idea of, 26–27
Petroleum industry, 155–156,
  161–165
Place, value of, 182, 188–191
Polar bear, as endangered, 111
Polyarchy, 115
Proudhon, P. J., 18

Rawls, John, 61, 73
Responsibility, principle of, 50, 62
Ricardo, David, 155
Rights
  to an adequate environment, 48–49

basic vs. nonbasic, 51–55, 64
  to climatic stability, 46–47, 49,
    63–64
  to development, 45, 55–64
  human, 54
  to security, 52–53
  to subsistence (*see* Equality)
Rule of law, 77–81

Schlosberg, David, 73
Scientific uncertainty, 29, 86n,
  146–147
Shue, Henry, xix, 51–55, 63
Social-ecological systems analysis,
  91–93, 96–103, 112–115
Stern Report, vii
Stockholm Declaration, 49,
  68
Survival vs. luxury emissions, xviii,
  7–8, 45–46, 51, 58–59

Technoscience, 136
Timber industry, 106–108, 156
Tragedy of the commons. *See*
  Collective action
Treaties
  Antarctic Treaty, 9–10
  Montreal protocol, 127
  Moon Treaty, xvii, 12–13
Twentieth Century Fund, 162–163,
  165

UN Conference on Environment and
  Development, 68
UN Conference on the Human
  Environment. *See* Stockholm
  Declaration
UN Convention on the Law of
  the Sea, 8–11, 144. *See also*
  Treaties
United Nations Charter, 72, 78
United Nations Framework
  Convention on Climate Change
  (UNFCCC)
  policy goals, xi–xiii, 43, 47, 56–57,
    65
  ratification of, 5, 70

Urban-natural hybrid, or *urbanatura*,
    xxi, 121, 28–138, 132
  as Second Creation, 134–136
Urban sprawl, 157–162

Virilio, Paul, 130–131
Vital interests, principle of, 51,
    53–54

Wildfire
  boreal forests, effects upon, 105–106,
    112
  management policies, 107–108, 112

Young, Iris Marion, 74